Picturing Animals and Plants in Early Modern China and Japan

Picturing Animals and Plants in Early Modern China and Japan

Innovation, Experiments, and Anxieties

Edited by
Fan Lin and Doreen Mueller

Routledge
Taylor & Francis Group
LONDON AND NEW YORK

The publication of this book is made possible by grants from the Isaac Alfred Ailion Foundation, the Heinz Kaempfer Fund, and Leiden University Institute for Area Studies (LIAS).

ailion foundation

Heinz Kaempfer Fund

First published in 2025 by Amsterdam University Press Ltd.

Published 2025 by Routledge
4 Park Square, Milton Park, Abingdon, Oxon OX14 4RN
605 Third Avenue, New York, NY 10158

Routledge is an imprint of the Taylor & Francis Group, an informa business

ISBN: 9789048559091 (hbk)
ISBN: 9781041184164 (pbk)
ISBN: 9781003701378 (ebk)
NUR 692

Cover illustration: Nie Huang 聶璜, "Whale" (Jingyu 井魚) in *Pictures of the Marine Creatures* (Haicuo tu 海錯圖). Album leaf. 1698. Palace Museum, Beijing.

Cover design: Coördesign, Leiden

DOI: 10.5117/9789048559091

Table of Contents

Acknowledgements

This book started as a workshop "Curious and Scientific Things – Seeing and Knowing in Early Modern East Asia (1700-1900)" held at Leiden University in October 2018 and sponsored by Leiden University's Asian Modernities and Traditions Research Fund. We are grateful to all participants, presenters, and discussants who contributed to this workshop. Their constructive feedback encouraged us to progress towards a book project.

During the process of editing this book, we have received tremendous support from many institutions, colleagues, and friends. We thank Leiden University's Institute for Area Studies (LIAS) for providing research funding in 2022 and 2023; the open access of this book was made possible thanks to the generous support of the Isaac Alfred Ailion Foundation, a LIAS Publication Grant, and the Heinz Kaempfer Fund. At various stages, we received valuable advice from Victoria Blud, Marc Buijnsters, Anne Gerritsen, Hilde De Weerdt, Marc Gilbert, Lianyu Jin, Ivo Smits, Nira Wickramasinghe, and Guanmian Xu. We are also grateful for the meticulous and insightful comments from the anonymous reviewers. Our copy editor Anna Beerens patiently read through our manuscript twice and offered numerous comments. Last but not the least, we thank our editor Inge Klompmakers and her team for providing unwavering support with their professional expertise and heartfelt encouragement.

List of Figures

Introduction

1. Singing Frogs: Approaches to Registering Animals in The *Nihon Sankai Meisan zue*

2. Tea Harvesting at Uji. Repackaging Uji as a Productive Place

3. Disciplined Objects? Wood panels from the Kew Collections

4. The Return of the Elephants. A Social History of Elephant Watching in Early Modern China

5. A Pair of Camels in Edo Japan. Representation and Discourse

6. *Pictures of Sea Fish* (*Haiyu tu*) and Knowledge of Nature in Eighteenth-Century China

7. Treatise (*pu*) versus Illustration (*tu*). The Absence and Presence of Illustrations in *Pulu* Writings on Chinese Nature Studies

Introduction

Fan Lin and Doreen Mueller

In early modern China and Japan, the meanings of animals and plants moved beyond the traditional frameworks of auspiciousness and usefulness. The chapters in this volume consider the early modern period as a critical phase for bringing together these notions with new empirical and epistemological concerns, political and social agendas, and cultural interests. They examine how the interactions between established and emerging media such as printed encyclopaedias, monographs on specialised topics, and images in circulation, allowed new meanings to be generated, and, within this process, how scholars, professional painters, and publishers became more conscious of their role in shaping the ways in which information about animals and plants were organised and represented.

To start with an example, let us consider an illustration of three *matsutake* mushrooms produced by the shogunal retainer and amateur naturalist Mōri Baien (1798–1851) in a volume on mushrooms dated 1836 (Figure 0.1).[1] Mōri's illustration is a statement of empirical method as he labelled it "drawn from life" (Chin. *xiezhen*, Jap. *shinsha*), and also expresses commercial concerns. Mōri claimed that he collected, trimmed, and drew the mushrooms on the third day of the ninth month. Accompanied by references to a Southern Song (1127–1279) treatise on mushrooms, a well-known Joseon (1392–1897) *materia medica*, and a presumed Kamakura Period (1192–1333) *waka* poem attributed to Fujiwara Teika (1162–1241), the drawing acquired an aura of scholarly authority.[2] His entry further describes selected local variations of *matsutake* mushrooms, assessing their appearance and taste, followed by information about their trade in Osaka and Kyoto. He dropped the line, "pick up anytime," from the Southern Song treatise, concretising with the assertion that they "grow plentifully in the ninth and tenth months." At the same time, he also fabricated the attribution of the poem to the renowned medieval

1 Mōri Baien, *Baien's Treatise on Mushrooms* (Baien Kinpu), 1836 (illustrated manuscript, National Diet Library, Tokyo), p. 22.

2 Baien cited *The Treatise on Mushrooms* (Junpu) by Chen Renyu (b. 1212) and *A Precious Mirror of Eastern Medicine* (Dongui Bogam).

Fan Lin and Doreen Mueller (eds), *Picturing Animals and Plants in Early Modern China and Japan. Innovation, Experiments, and Anxieties.* Taylor & Francis Group, 2025
DOI: 10.5117/9789048559091_INTRO

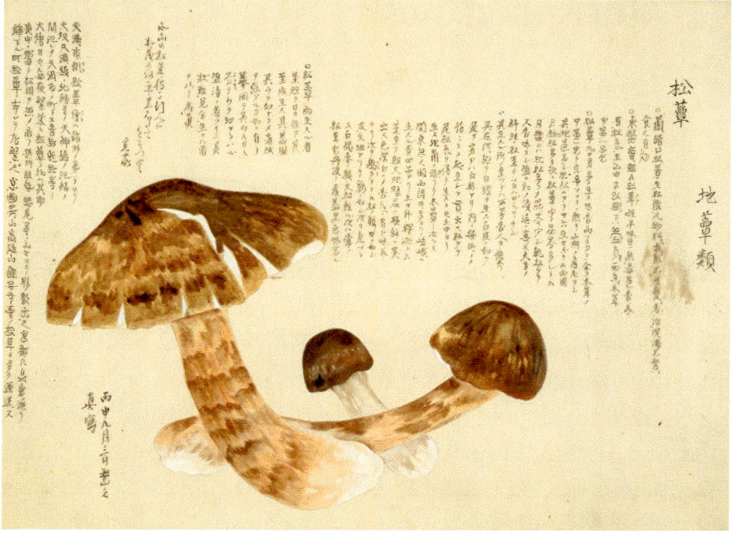

Figure 0.1: Mori Baien, Matsutake mushrooms, in *Baien's Treatise on Mushrooms* (*Baien kinpu*) (manuscript 1836, National DIET Library, Tokyo), p. 22.

court poet Fujiwara Teika.[3] Mōri's crafted combination of illustration and written caption shows how scholars attempted to establish authority by situating their empirical observations and commercial interests within a canonical cultural and historical framework.

Moving beyond Auspiciousness

Before the early modern period, the concepts of animals and plants were not a given. They were raised in one of the core Confucian classics, *The Rites of Zhou* (Zhouli), during the Western Han (206 BCE–9CE) period. *Dongwu* literally means "moving things" and it has been consistently applied to animals; *zhiwu* means "planted things," and thus plants.[4] In *The Rites of Zhou*, *dongwu* and *zhiwu* were coded to create hierarchies in the ordered world by distinguishing the tributary produce from five types of soils:

3 The transcription of the poem is as follows: "People who go hunting for pine mushrooms in the Northern Hills (of Kyoto) should first pick the shoots of the water pepper (*tade*) from the bed of Kamo River." The poem refers to the inherent dangers of picking mushrooms by mentioning the water pepper, which was used as an antidote for poisonous mushrooms. The poem was not included in anthologies of poetry by Fujiwara Teika and is therefore most likely a fabrication by Mori. We are grateful for the fruitful discussions with colleagues at Leiden University, especially Marc Buijnsters and Ivo Smits.

4 Zhi also means 'standing' or 'immovable.' See Martina Siebert, "Klassen und Hierarchien, Kontrastpaare und Toposgruppen: Formen struktureller Eroberung und literarischer Vereinnahmung der Tierwelt im alten China," *Zeitschrift Der Deutschen Morgenländischen Gesellschaft* 162 (2012: 1): 172.

> To use the methods of reckoning soil to discern the produce from five categories of land:
> The first is mountains and forests: Regarding animals, it is appropriate for hairy animals; regarding plants, it is appropriate for nuts; its people are hairy and rectangular.
> The second is rivers and marshes: Regarding animals, it is appropriate for scaly animals; regarding plants, it is appropriate for aquatic plants; its people are black and fat.
> The third is hills and slopes: Regarding animals, it is appropriate for feathered animals; regarding plants, it is appropriate for nut trees; its people are round and long.
> The fourth is banks and plains: Regarding animals, it is appropriate for armoured animals; regarding plants, it is appropriate for beans; its people are white and lean.
> The fifth is lowlands and marshes: Regarding animals, it is appropriate for light-coloured animals; regarding plants, it is appropriate for reed grasses, its people are fleshy and short.[5]

In the above passage, *dongwu* and *zhiwu* are categorically independent from humans.[6] However, these terms were rarely used as epistemological frameworks until the twentieth century in East Asia.[7] The earliest extant dictionary, *Approaching Elegance* (Erya) (Third century BCE), covers both animals or plants in manifold categories, such as herbaceous plants, trees, insects, fish, birds, beasts, and domesticated animals. In the nineteenth century, the terms *dongwu/dōbutsu* and *zhiwu/shokubutsu* were employed to translate Western works and gained strong scientific implications. Fukuzawa Yukichi (1834–1901), the most influential thinker in Meiji (1868–1912) Japan, adopted *dōbutsu* as a neologism in response to Western science.[8] Throughout the history of East Asia, the categories in *Approaching Elegance* were commonly used in treatises and encyclopaedias, shaping how people understood the worlds of animals and plants. The oft-cited encyclopaedia, *Illustrated Compendium of the Three Powers* (Sancai tuhui), adhered to the governing logic of *Approaching*

5 Jia Gongyan, *Zhouli zhushu*, in *Chongkan Songben shisanjing zhushu* (Nanchang: Nanchang Prefectural Academy, 1815), p. 150.

6 Roel Sterckx, *The Animal and the Daemon in Early China* (Albany, NY: State University of New York Press, 2002), p. 19.

7 Sterckx, *The Animal and the Daemon in Early China*, pp. 16–21. Gregory M. Pflugfelder and Brett L. Walker, *JAPANimals: History and Culture in Japan's Animals Life* (Ann Arbor, MI: Center for Japanese Studies, University of Michigan, 2005), p. xii.

8 Roel Sterckx, Martina Siebert, and Dagmar Schäfer eds., *Animals Through Chinese History: Earliest Times to 1911* (Cambridge: Cambridge University Press, 2018), p. 1. Ian Miller, "Didactic Nature: Exhibiting Nation and Empire at the Ueno Zoological Gardens" in *JAPANimals: History and Culture in Japan's Animals Life*, p. 297. Pflugfelder, preface, *JAPANimals: History and Culture in Japan's Animals Life*, preface, p. xii.

Elegance, and expanded existing categories (herbaceous plants and trees, and birds and beasts) into many subcategories.[9]

More specifically, a multitude of animals and plants were deeply intertwined with the system of auspicious referents and integral to everyday life. They were placed on buildings, doors, gates, clothing, carriages, utensils, and accessories. Paintings of almost every plant and animal were pleasing to the eye but at the same time they were taken to be talismanic objects endowed with the power to ward off evil elements. There were dragons, *qilin* (*kirin*), gibbons, goats, cranes, magpies, bamboo, pine tree, peach trees, chrysanthemums, peonies, plums, and lotus; the list is endless. The concept of auspiciousness was pertinent in East Asian cultures, but the meanings and connotations of auspicious animals and plants were not timeless and placeless.

In the early modern period, the auspiciousness attached to animals and plants underwent an epistemic transformation. Rather than being an omniscient force orienting people's lives, urban and rural dwellers started to assume a dominant position instrumentalising auspicious animals and plants within the construction of their lived realities. The legendary story that Cao Buxing's (third century CE) painting of a vermillion dragon can produce rain is often interpreted as praise for his miraculous painting skills, but it also points to the magical power of the dragon to interact with the human world.[10] In his unpublished picture book, the Edo painter Katsushika Hokusai (1760–1849) drew the dragon through a mundane, empirical lens by comparing its elongated shape and surface textures with a crocodile, an octopus, and other sea creatures.[11] In the case of elephants, although they were generally understood as sacred animals in the Buddhist cultural sphere, in the imperial vision of ethnic communities in the southwest of the Qing empire, the elephants served as local resources facilitating transportation (Figure 0.2).

These changes did not mean that auspicious animals or plants completely lost their magical power or symbolic connotations, but rather that they underwent a process of vernacularisation and became a backdrop to urban life. Their growing closeness to human realities amplified their perceived auspiciousness. The pine tree, for example, is a common symbol in East Asia connoting longevity, integrity, peace, and benevolent rule. When they appeared in the background of harvesting and fishing in paintings or illustrations, however, they were often used as practical narrative props framing scenes, but their latent presence was simultaneously a potent agent in eliciting positive affective responses from viewers (Chapter 1, Chapter 2).

9 Wang Qi and Wang Siyi, *Sancai tuhui* (Shanghai: Shanghai guji chubanshe, 2002); Terajima Ryōan, *Wakan Sansai zue* (Publisher Unknown: 1712).

10 Zhang Yanyuan, *Lidai minghua ji* (Shanghai: Shanghai renmin meishu chubanshe, 1964), pp. 89–90.

11 For Hokusai's illustrations, see Sarah E. Thompson, *Hokusai's Lost Manga* (Boston, MA: Museum of Fine Arts Boston, 2016), pp. 43–45.

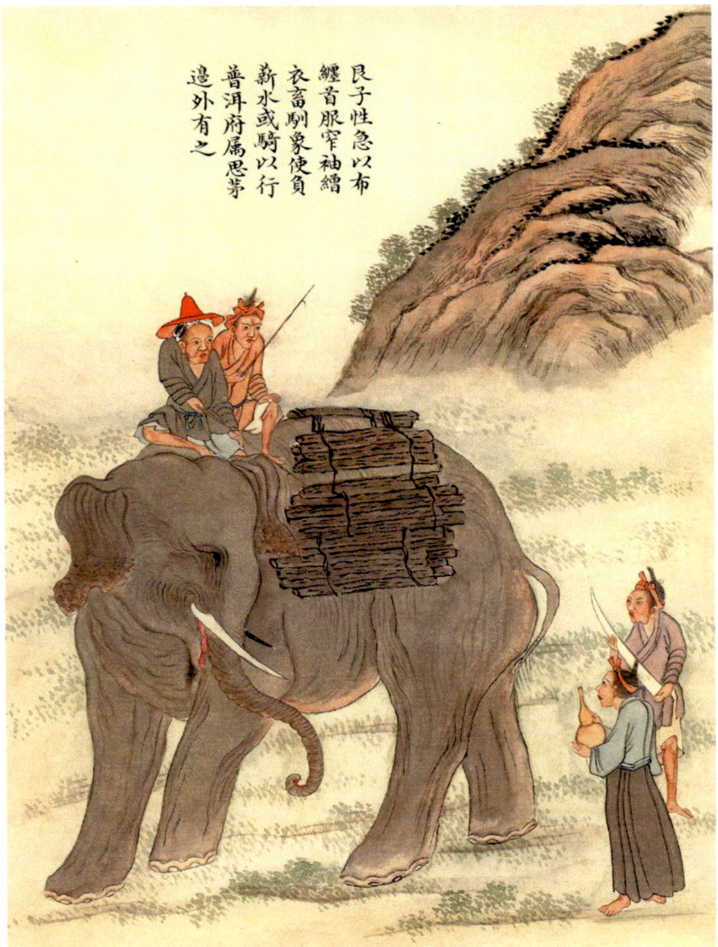

Figure 0.2: Chuai Zhenyu, An illustrated guide to an Elephant with Two Mahouts, in *An Illustrated Guide to the Ethnic Groups in Yunnan Province* (Diansheng yiren tushuo). (Beijing: Zhongguo shehui kexue chubanshe, 2009), p. 85.

The seven chapters in this volume draw attention to several aspects arising from the process of image making in early China and Japan: the role of images in foregrounding the usefulness of animals and plants; making knowledge about exotic animals and plants public; and addressing scholars' anxieties concerning the epistemic capacity of images and texts.

Useful Images and Useful Things

Useful animals and plants in this volume are taken as those documented in books on agriculture and *materia medica*. Foundational agricultural works, such as *Fan Shengzhi's Book* (Fan Shengzhi shu, first century BCE) and *Essential Ways of Securing the Livelihood of the Common People* (Qimin yaoshu, sixth century CE) gave weight to practical knowledge about growing and harvesting plants and animals that could benefit people. They were conceptualised as useful things, such as staple food, vegetables, cash crops, and livestock. However, in the early modern period, usefulness became increasingly localised and gained added values. The Confucian scholar Miyazaki Yasusada's (1623–1697) *Compendium on Agriculture* (Nōgyō zensho, 1697) was based on the Ming agricultural encyclopaedia, *Complete Treatise on Agricultural Administration* (Nongzheng quanshu, 1639), but especially promoted the practical learning of local crops in Japan. He drew attention to the benefits of closely observing and registering the appearance of cash crops such as rice or sweet potato, including their careful cultivation according to the local climate.[12] In images, things are never just things, but can be endowed with social relevance. In *The Works of Heaven and the Inception of Things* (Tiangong kaiwu, 1637) as well as the above-mentioned *Compendium on Agriculture*, illustrations established a harmonious relationship between people and the environment in the production process.[13] This was taken further in the following decades, as illustrations showed this relationship through the gaze of onlookers from different backgrounds.[14] Images foregrounded these added values, to suggest the desirability of things and to nurture a sense of shared community. Commercial publishers presented fishing on Kamo River as a local event, a spectacle that invited urban dwellers to watch. They framed the usefulness of river fish in larger economic and social contexts (Chapter 1). This shows that this sense of shared community was in the direct interest of publishers who wanted to appeal to broader, urban audiences. Similarly, in painting, tea plantations became a scenic site in Kano school painting scrolls, evoking a sense of co-productivity and harmony between humans and plants, with the growing tea plants mirroring the happy state of women, children, and families (Chapter 2).

Katō Chikusai's (1818–1886) wood panels shifted the notion of usefulness away from its perceived harmonious relation with humans. Commissioned by Koishikawa Botanical Garden at Tokyo Imperial University, a modern academic institution, Katō's works promoted Japanese woods as commercially viable building materials at the international stage by demonstrating usefulness with wood samples and employing universal Latin nomenclature. Although scholars interested in *materia medica* in the early nineteenth century used ink rubbings to represent plant specimens, Katō's wood samples were an even more direct and effective

12 Christine Guth, *Craft Culture in Early Modern Japan: Materials, Makers, and Mastery* (Berkeley, CA: University of California Press, 2021), p. 41.

13 Dagmar Schäfer, *The Crafting of the 10000 Things: Knowledge and Technology in Seventeenth-Century China* (Chicago, IL, and London: University of Chicago Press, 2011), pp. 138–156.

14 Robert Goree, *Printing Landmarks: Popular Geography and Meisho Zue in Late Tokugawa Japan* (Cambridge, MA: Harvard University Asia Center, 2020).

method for embodying the represented plants.[15] In so doing, Katō's works pointed to a new mode of visualising usefulness in the scientific rhetoric. In other words, they extended the notion of usefulness from trees to the medium of representation.

Exotic Animals and Plants: Making Them Public

With increasing interactions between East Asia and the global world, exotic animals and plants became more present in the public consciousness, but their "foreignness" resisted an easy packaging using existing taxonomies. Take *The Compendium of Materia Medica* (Bencao gangmu) as an example in which exotic animals and plants were incorporated into the existing categories of minerals, herbs, grains, vegetables, fruits, trees, insects, animals with scales, animals with shells, birds, and beasts. Under the category of fruits, thirty-one types of exotic fruits were put under the new separate subcategory of "foreign fruits" (*yiguo*). Instead of creating completely new ways to conceptualise them, the foreignness was tamed by applying existing knowledge systems and practices to them. The foreign fruits in *The Compendium of Materia Medica* continued to be described with the standard language corresponding to the systems of Five Phases, Five Tastes, and Cooling or Warming effects. Palm dates from Persia, for example, were "sweet, warming, and non-toxic"; they were used to "help digestion, stop coughing, cure depletion, and make people happy, and they cause no harm if taken for a long period."[16]

The epistemic logic of *materia medica* required viewing an animal or plant as an assemblage of constituent parts that were deemed beneficial to humans. For example, a camel was medically useful only when its hump, meat, milk, gallstones, fur, or dung became medical ingredients.[17] Although the illustrated Sino-Japanese encyclopaedia *Illustrations of Three Powers in Japan and China* (Wakan sansai zue) citing *Bencao gangmu* and its Chinese version, *Illustrations of Three Powers*, were used as sources to describe the appearance of the camel, the Edo illustrations of camels largely disregarded the medical functions of their body parts. In the public entertainment circuit, camels were paraded as exotic animals. This created possibilities for generating alternative meanings that transcended the *materia medica* concept of usefulness. For example, scholars attending the camel shows commissioned painters skilled in empirical painting methods to capture their docile demeanor, which was emblematic of a carefree attitude and conjugal harmony.

15 Maki Fukuoka, *The Premise of Fidelity Science, Visuality, and Representing the Real in Nineteenth-Century Japan* (Stanford, CA: Stanford University Press, 2012).

16 Li Shizhen, *Bencao gangmu*, in *Yingyin Wenyuange Siku quanshu*, vol. 772–774 (Taipei: Taiwan shangwu yinshuguan, 1983), pp. 31.23b–24b.

17 Li Shizhen, *Bencao gangmu*, pp. 50b.34a–35b.

Giving animals "distinctly human features, virtues, or vices or, likewise, to render botanical or geological motifs with personified features" takes the collective human as the centre, but these Edo camel portraits also show them as sentient beings with distinct personalities (Chapter 5).[18] Similarly, washing the imported elephants in the moats near the Ming imperial palace was also an urban spectacle and, in combination with its Buddhist association, it gained the connotation of purification of the mind (Chapter 4).

A Scholar's Problem: Epistemic Anxieties about Image and Text

The definition of early modernity in East Asia is not uniform, but scholars have reached a consensus that it features growing urbanisation, commodification, literacy, and exploitation of natural resources.[19] Mark Elvin, Søren Clausen, and Antonia Finnane have argued that the Song dynasty (960–1279) constituted the early phase of this profound transformation; more commonly, scholars associate Ming (1368–1644) and Qing (1644–1911) China, Joseon Korea (1392–1897), and Tokugawa (1615–1868) Japan with the traits of early modernity.[20] The chapters in this volume cover the *longue durée* from 1600–1900, a period that witnessed these social and economic changes at an exponential scale. Meanwhile, the scholarly elite continued to be the key producers of knowledge, and increasingly educated merchants, professional painters, and illustrators also became active participants.

While scholars still enjoyed academic and social prestige, they started to become conscious of nascent societal changes. In the process of studying and presenting their knowledge, they rethought the roles of images and texts, thereby becoming critical of entrenched practices and ideas based on centuries of traditions of representing knowledge in *materia medica* and other encyclopaedic media. While a seemingly visual turn was happening in early modern East Asia, the educated elites were aware of the precarious status of images as carriers of knowledge. Already in the Song dynasty, scholars who authored treatises (*pulu*) made use of illustrations drawn by professional painters as visual aids in the process of collecting information

18 Jerome Silbergeld, "Trading Places: An Introduction to Zoomorphism and Anthropomorphism in Chinese Art," in Jerome Silbergeld and Eugene Wang, eds., *The Zoomorphic Imagination in Chinese Art and Culture* (Honolulu: University of Hawaii Press, 2016), p. 1.

19 Søren Clausen, *Early Modern China—A Preliminary Postmortem* (Aarhus: Aarhus Universitet, 2000); Brett L. Walker, *JAPANimals: History and Culture in Japan's Animals Life*, introduction, pp. 1–20.

20 Clausen, *Early Modern China—A Preliminary Postmortem*; Antonia Finnane, "Folding Fans and Early Modern Mirrors," in Martin J. Powers, Katherine R. Tsiang, and Dana Arnold, eds., *A Companion to Chinese Art*, (Chichester: John Wiley & Sons, 2015), pp. 396–397. Craig Clunas, *Pictures and Visuality in Early Modern China* (London: Reaktion Books, 1997), pp. 9–10.

but left them out in the final *pulu* work (Chapter 7). At the same time, officials like Han Liangqing (d. 1740) saw a new role for images in materialising a narrative taxonomic order when he commissioned a professional painter in Guangdong province to carefully document the appearance, names, and ways of cooking a variety of local fish (Chapter 6). The period between 1600 and 1900 was a dynamic time of negotiation in which scholars explored new inroads and reconsidered the role of images in shaping knowledge. Within this process, images that were based on direct observation became new sources of authoritative knowledge. Although access to animals and plants became more "democratic" as, for example, exotic animals, such as camels, were exhibited in public shows, the right to derive knowledge from them still largely hinged on the social standing of the image producers as well as their epistemic and cultural capital (Chapter 5). In this sense, images labelled as empirical were shaped by social historical realities.

Although professional painters were conventionally understood to be commissioned to produce paintings and illustrations, the chapters in this volume bring attention to their agency in and contribution to knowledge production. The painters discussed in this volume represent a broad spectrum, ranging from notable master painters and illustrators to lesser-known ones. In the case of the *Pictures of Sea Fish* (Haiyu tu) and some *pulu*, the names of the painters and illustrators were overshadowed by the scholars who orchestrated the projects, but their contribution should be seen as part of the negotiation process between the emerging empirical practices and the existing auspicious meanings of animals and plants (Chapter 6 and Chapter 7). Even when the drawings were not included in the final *pulu*, there were cases when they constituted a crucial step in understanding the physical form of the plants (Chapter 7). Likewise, scholars were inspired by the urban spectacle of watching elephants, but it was the painters who visualised this invented tradition (Chapter 4). Painters of camels and elephants, for example, exercised their agency not just by capturing their physical appearance, but also by projecting human traits onto the animals. At the same time, images of plants catered to nascent economic concerns. In Kano school paintings, the image of Uji as a famous place associated with aristocratic culture shifted towards a productive place where high-quality green tea was grown as a cash crop (Chapter 2). At the end of the time frame of this volume, Katō Chikusai (1818–1886) took the role of the painter in an even more practical direction by integrating wood samples into his paintings serving the economic interests of the Meiji state (Chapter 3).

Educated merchants joined this process as impresarios of public shows, publishers of compendiums, and authors themselves. On one hand, they continued to support the printing of Confucian classics related texts; on the other hand, they also produced writings through more unconventional channels, which allowed them to chart out the new potentials and possibilities of envisioning the known and the unknown natural world. The latter took a more malleable form, and sometimes existed as an open-ended project spanning a few centuries. Take the categories of "Agriculture and Mulberry," "Flowers, Fruits, Bamboos, and Trees," and "Beasts and Domesticated Animals" in the encyclopaedia *Comprehensive Compendium in the Forest of Affairs* (*Shilin guangji*). It was first compiled by Chen Yuanliang (active in the second half of the thirteenth century) in the late Southern Song (1127–1279) but was edited multiple times in the following centuries. By the Ming, various editions were in circulation in East Asia, and, in comparison with each other, a shrewd reader would notice that their taxonomical structures were constantly being updated and their contents revised and supplemented.[21] As Lucille Chia and Cynthia Brokaw have observed, the rivalry between various editions lay in a pursuit of lower printing costs; the authenticity of the original contents and authorship were not a serious concern of the printing workshops or their readers.[22] In eighteenth- and nineteenth-century Japan, illustrated travel guides to famous places (*meisho zue*) and encyclopaedias pertaining to animals and plants were illustrated with increasingly elaborate double-pages that gave weight to visual experiences; the power of close observation went side by side with epistemic query and commercial interests.[23] Merchants in Kyoto and Osaka cross-referenced canonical works and updated them with contemporary information taken from their own observations and from illustrations of rural life (Chapter 1).

The democratisation of knowledge production challenged the uncontested status of writing as the foundation of scholarly identities. As a rule, scholars developed their writing skills to establish their career, including passing exams, expressing themselves, and forming networks. At the same time, both governments and scholars increasingly found illustrations in printed books useful for interpreting the structure of the cosmos, geomancy, architecture, *materia medica*, geography, and engineering works. Although often accompanied by texts, these illustrations were more than just visual aids. As some of them acquired greater mimetic authority as things in themselves, they started challenging the established status of writing (Chapters 3 and 5). Anxieties about the effectiveness of illustrations, particularly printed illustrations, came from the question of their role as appropriate scholarly

21 Etienne Balazs and Yves Hervouet, *A Sung Bibliography* (Bibliographie des Sung), p. 328. Chen Yuanliang, *Shilin guangji*, in *Hekeben leishu jicheng* (Shanghai: Shanghai guji, 1990), vol. 1.

22 Brokaw, "Mashaben: Commercial Publishing in Jianyang from the Song to the Ming," in Paul Smith, Richard Von Glahn, Peter K. Bol, Lucille Chia, and Angela Ki Che Leung, eds., *The Song-Yuan-Ming Transition in Chinese History* (Cambridge, MA: Harvard University Asia Center, 2003), pp. 284–328; Lucille Chia, *The Commercial Publishers of Jianyang, Fujian (11th–17th Centuries)* (Cambridge, MA: Harvard University Asia Center 2002); Yuming He, *Home and the World: Editing the Glorious Ming in Woodblock-Printed Books of the Sixteenth and Seventeenth Centuries* (Boston, MA: Harvard University Asia Center, 2013), pp. 1–17.

23 Goree, *Printing Landmarks*.

tools to present knowledge. This might explain why Cai Xiang (1012–1067) did not include illustrations in his final printed project, despite having a professional painter draw various types of lychees (Chapter 7).

Towards the end of the early modern period, this tension was no longer a major problem. The scholar official Han Liangqing confidently used hand-painted drawings rather than text in his album describing local fish in Guangdong province. Starting in the late eighteenth century, and especially in Japan, naturalist scholars devised images to identify different species of the same animals or plants. The growing accuracy of images was driven by the practical need to make sense of the ever-expanding knowledge of nature.[24] Scholars who pursued knowledge of the medicinal properties of animals and plants, such as Kurimoto Tanshū (Masayoshi, 1756–1834), Hiraga Gennai (1729–1780), and Iwasaki Tsunemasa (1786–1842), took images to the centre stage.[25] In the same vein, Mōri Baien, mentioned above, also took images seriously in his personal pursuit of knowledge about mushrooms as well as other subjects in his mammoth natural history project. More importantly, he not only referenced the canonical knowledge but made it local and personal by including his own observations and experiences. In the case of mushrooms, he claimed that each item was either harvested by himself or given to him by local people or friends, and thus his "drawn from life" approach gave them an aura of authenticity.

Chapter Abstracts

The first chapter considers how printed reference works in early modern Japan like the *Illustrated Guide to Famous Products of Land and Sea in Japan* (Nihon sankai meisan zue), first published in 1799, represented nature as an assemblage of objects while also highlighting its close connection with local customs and practices of manufacturing goods. It discusses how eclectic modes of representing nature in image and text in this illustrated guide indexed a multitude of ways to expand knowledge and to make it reliable and useful. The illustrated guide has a preface by Kimura Kenkadō (1736–1802), a successful brewer, rice wine merchant, and industrious amateur scholar. The guide in its entirety was probably put together by several members of Kenkadō's amateur scholar circle in and around Osaka. Its eclectic style shows the intersection of their manifold interests: the empirical study of nature as well as historical and literati pursuits.

24 Federico Marcon, *The Knowledge of Nature and the Nature of Knowledge in Early Modern Japan* (Chicago, IL: The University of Chicago Press, 2015), pp. 248–250.

25 Marcon, "All Creatures Great and Small: Tokugawa Japan and Its Animals," p. 32.

The second chapter investigates the production and dissemination of knowledge about a particular region in early modern Japan, namely, late-seventeenth- century to early-eighteenth-century Uji. The case study of Uji shows how traditional knowledge was being negotiated with new economic and cultural interests. Taking the scroll *Tea Harvesting at Uji* by Kanō Tansetsu as its focus, this chapter argues that the pictorial language of Uji was manipulated to promote its image as a productive tea region peacefully governed by the Tokugawa Shogunate.

The third chapter analyses a set of twenty-six wooden panels currently held at Kew's Economic Botany Collection. Katō Chikusai produced the set at Koishikawa Botanical Garden in Tokyo in 1878. By taking seriously the methodological and discursive presuppositions that are often and easily overlooked in studies of early Meiji visual objects, this chapter reorients the perspective to one that privileges the context of the production site of the wood panels over their current location. Through historically informed analyses, this chapter highlights how these objects carry two constitutive values — economic and epistemic — of early Meiji Japan.

The fourth chapter examines the social history of watching elephants in late imperial China. Elephants were native to early China, but they had migrated to the southwestern margin of the country and Southeast Asia by the tenth century. After the Song dynasty, they were brought back to the capital as tributary animals or diplomatic gifts. They entered the public arena by, for example, being used in imperial parades. In the Ming dynasty, elephant washing in the city moat outside the imperial palace in early summer evolved into a captivating urban spectacle. This phenomenon instigated the creation of various artworks but also took on additional layers of social significance.

The fifth chapter examines the striking reception of a pair of single-humped camels in Edo Japan. The Dutch East India Company brought the camels to Nagasaki in 1821 as diplomatic gifts to the Tokugawa shogun. Rejected by the shogun, the camels were displayed in public shows around the country for years, attracting enormous attention. The popularity of the shows caused an explosion of camel-related broadsheets and *ukiyo-e* prints. The camel shows not only provided a wide audience with information, both genuine and fabricated, but also inspired scholars, intellectuals, writers, and painters to produce novels, verses, songs, essays, and paintings. The camels' reception in Edo society generated a wealth of connotations concerning camels, which also reflected people's general view of animals. Particularly noteworthy is that the Buddhist idea of the non-duality of humans and animals reverberated in the work of naturalistic painters such as Maruyama Ōshin (1790–1838).

The sixth chapter examines a relatively unknown Chinese album *Pictures of Sea Fish* by Han Liangqing. It depicts more than 130 species of sea fish. It is a world away from the traditional artistic representation of fish and has more in common with the scientific investigations of natural history. This chapter examines this album, and subsequently situates it in the contexts of nature studies and visual culture in early modern China. As the first research on this album, the chapter sheds light on the role of scientific illustrations within the formation of knowledge of the natural environment in China in the eighteenth century.

The seventh chapter examines the complex relationship between text and image in *pulu* writing about animals and plants. *Pulu* began to thrive as a genre in the twelfth century, a time when bibliographer and historian Zheng Qiao coined the notion of image (*tu*) and descriptive text (*pu*) as mutually supportive means for packaging knowledge. *Pulu* are often expected to be illustrated but, in fact, this is rarely the case. The chapter begins with brief probes into the history of the bibliographical classification of *pu*- and *tu*- type books and some reflections on the appearance of *tu* in the titles of works. It then explores examples of non-illustrated and illustrated *pulu* and the circumstances and scholarly attitudes that informed the choice of one or the other.

Glossary

Baien kinpu	梅園菌譜
Bencao gangmu	本草綱目
Cai Xiang (1012–1067)	蔡襄
Cao Buxing (third century)	曹不興
Chen Renyu (b. 1212)	陳仁玉
Chen Yuanliang (thirteenth century)	陳元靚
dongwu (Chin.)/*dōbutsu* (Jap.)	動物/動物
Erya	爾雅
Fan Shengzhi shu	氾勝之書
Fujiwara Teika (1162–1241)	藤原定家
Fukuzawa Yukichi (1834–1901)	福澤諭吉
Guangdong	廣東
Haiyu tu	海魚圖
Han Liangqing (d. 1740)	韓良卿
Hiraga Gennai (1729–1780)	平賀源内
Iwasaki Tsunemasa (1786–1842)	岩崎常正
Junpu	菌譜
Kamo River	鴨川
Katō Chikusai (1818–1886)	加藤竹斎
Kanō Tansetsu (1654–1713)	狩野探雪
Katsushika Hokusai (1760–1849)	葛飾北斎
Kimura Kenkadō (1736–1802)	木村蒹葭堂

Koishikawa	小石川
Kurimoto Tanshū (Masayoshi, 1756–1834)	栗本丹洲
Maruyama Ōshin (1790–1838)	円山応震
meisho zue	名所図会
Miyazaki Yasusada (1623–1697)	宮崎安貞
Mōri Baien (1798–1851)	毛利梅園
Nihon sankai meisan zue	日本山海名産図絵
Nōgyō zensho	農業全書
Nongzheng quanshu	農政全書
Owari Domain	尾張国
pulu	譜錄
qilin (Chin.)/*kirin* (Jpn.)	麒麟/麒麟
Qimin yaoshu	齊民要術
Sancai tuhui	三才圖會
Shilin guangji	事林廣記
shinsha	眞寫
Shōhyakusha	嘗百社
Tiangong kaiwu	天工開物
tu (Chin.)/*zu* (Jap.)	圖/図
Uji	宇治
Wakan sansai zue	和漢三才図会
xiezhen	寫真
yiguo	異果
zhiwu (China)/*shokubutsu*	植物/植物
Zhouli	周禮

Bibliography

Primary Sources

Chen Yuanliang 陳元靚. *Shilin guangji* 事林廣記. In *Hekeben leishu jicheng* 和刻本類書集成, vol.1. Shanghai: Shanghai guji, 1990.

Jia Gongyan. *Zhouli zhushu* 周禮注疏. In Ruan Yuan 阮元 (1764–1849) compiled, *Chongkan Songben shisanjing zhushu* 重刊宋本. Taipei: Yiwen yinshuguan, 1965 (1815 Nanchang fu xue edition); incorporated in Scripta Sinica, Hanji dianzi wenxian ziliaoku 漢籍電子文獻資料庫 (hanji.sinica.edu.tw).

Li Shizhen 李時珍. *Bencao gangmu* 本草綱目. In *Yingyin Wenyuange Siku quanshu* 景印文淵閣四庫全書, vol. 772–774. Taipei: Taiwan shangwu yinshuguan, 1983.

Mōri Baien (1798–1851) 毛利梅園. *Baien kinpu* 梅園菌譜. Illustrated manuscript, 1836.

Ruan Yuan 阮元 ed. *Chongkan Songben shisanjing zhushu* 重刊宋本十三經注疏. Nanchang: Nanchang Prefectural Academy, 1815.

Terajima Ryōan 寺島良安. *Wakan Sansai zue* 和漢三才図会. Publisher Unknown, 1712.

Wang Qi 王圻 and Wang Siyi 王思義. *Sancai tuhui* 三才圖會. Shanghai: Shanghai guji chubanshe, 2002.

Zhang Yanyuan 張彥遠. *Lidai Minghua ji* 歷代名畫記. Shanghai: Shanghai renmin meishu chubanshe, 1964.

Secondary Sources

Balazs, Étienne, and Yves Hervouet. *A Sung Bibliography* (*Bibliographie Des Sung*). Hong Kong: The Chinese University Press, 1978.

Brokaw, Cynthia. "Mashaben: Commercial Publishing in Jianyang from the Song to the Ming." In Paul Jakov Smith, Richard von Glahn, Peter K Bol, Lucille Chia, and Angela Ki Che Leung eds., *The Song-Yuan-Ming Transition in Chinese History*, Cambridge, MA: Harvard University Asia Center, 2003, pp. 284–328.

Chia, Lucille. *Printing for Profit: The Commercial Publishers of Jianyang, Fujian (11th–17th Centuries)*. Boston, MA: Harvard University Asia Center, 2003.

Chia, Lucille. *The Commercial Publishers of Jianyang, Fujian (11th–17th Centuries)*. Cambridge, MA: Harvard University Asia Center, 2002.

Chuai Zhenyu 揣振宇. *Diansheng yiren tushuo* 滇省夷人圖說. Beijing: Zhongguo shehui kexue chubanshe, 2009.

Clausen, Søren. *Early Modern China: A Preliminary Postmortem*. Volume 84 of Arbejdspapirer, grøn serie. Aarhus: Aarhus Universitet, 2000.

Clunas, Craig. *Pictures and Visuality in Early Modern China*. London: Reaktion Books, 1997.

Finnane, Antonia. "Folding Fans and Early Modern Mirrors." In Powers, Martin J. Powers, Katherine R. Tsiang, and Dana Arnold eds., *A Companion to Chinese Art*. Chichester: John Wiley & Sons, 2015, pp. 392–409.

Fukuoka, Maki. *The Premise of Fidelity Science, Visuality, and Representing the Real in Nineteenth-Century Japan*. Stanford, CA: Stanford University Press, 2012.

Goree, Robert. *Printing Landmarks: Popular Geography and Meisho Zue in Late Tokogawa Japan*. Cambridge, MA: Harvard University Asia Center, 2020.

Guth, Christine M. E. *Craft Culture in Early Modern Japan*. Berkeley, CA: University of California Press, 2021.

He, Yuming. *Home and the World: Editing the Glorious Ming in Woodblock-Printed Books of the Sixteenth and Seventeenth Centuries*. Boston, MA: Harvard University Asia Center, 2013.

Marcon, Federico. "All Creatures Great and Small: Tokugawa Japan and Its Animals." In Robert T. Singer and Kawai Masatomo eds., *The Life of Animals in Japanese Art*. Princeton, NJ: Princeton University Press, 2019, pp. 29–37.

Marcon, Federico. *The Knowledge of Nature and the Nature of Knowledge in Early Modern Japan*. Chicago, IL: The University of Chicago Press, 2015.

Miller, Ian. "Didactic Nature: Exhibiting Nation and Empire at the Ueno Zoological Gardens." In Gregory M. Pflugfelder and Brett L. Walker eds., *JAPANimals: History and Culture in Japan's Animal Life*. Ann Arbor, MI: Center for Japanese Studies, University of Michigan, 2005, pp. 273—314.

Pflugfelder, Gregory M., and Brett L. Walker eds. *JAPANimals: History and Culture in Japan's Animal Life*. Ann Arbor, MI: Center for Japanese Studies, University of Michigan, 2005.

Schäfer, Dagmar. *The Crafting of the 10,000 Things: Knowledge and Technology in Seventeenth-Century China*. Chicago, IL, and London: University of Chicago Press, 2011.

Siebert, Martina. "Klassen und Hierarchien, Kontrastpaare und Toposgruppen: Formen struktureller Eroberung und literarischer Vereinnahmung der Tierwelt im alten China," *Zeitschrift Der Deutschen Morgenländischen Gesellschaft* 162 (2012:1): 171–195.

Silbergeld, Jerome. "Trading Places: An Introduction to Zoomorphism and Anthropomorphism in Chinese Art." In Jerome Silbergeld and Eugene Wang eds., *The Zoomorphic Imagination in Chinese Art and Culture*. Honolulu: University of Hawaii Press, 2016, pp. 1–20.

Sterckx, Roel, Martina Siebert, and Dagmar Schäfer eds. *Animals through Chinese History: Earliest Times to 1911*. Cambridge: Cambridge University Press, 2018.

Sterckx, Roel. *The Animal and the Daemon in Early China*. Albany, NY: State University of New York Press, 2002.

Thompson, Sarah E. *Hokusai's Lost Manga*. Boston, MA: Museum of Fine Arts Boston, 2016.

1. Singing Frogs: Approaches to Registering Animals in *The Nihon Sankai Meisan zue*

Doreen Mueller

Abstract: This chapter considers how printed reference works in early modern Japan, with a focus on the *Nihon sankai meisan zue* (*Illustrated Guide to Famous Products of Land and Sea in Japan*, first published in 1799), represented nature as an assemblage of objects while also highlighting its close connection with local customs and practices of manufacturing goods. It discusses how the diverse modes of representing nature in image and text in this illustrated guide indexed a multitude of ways of expanding knowledge and of making it reliable and useful. Its eclectic style shows the intersection of manifold interests: the empirical study of nature as well as historical and literati pursuits.

Keywords: nature studies, *meisho zue* 名所図会, materia medica, encyclopaedia

Knowledge is shaped by the practices that produce it: one might study animals and plants as phenomena to be observed out there, as sets of objects to be isolated, collected, and recorded for close-up study, or one might emphasise their cultural and economic meanings as resources for human industry. In addition, these approaches need not be mutually exclusive. Printed reference works in early modern Japan, such as the *Illustrated Guide to Famous Products of Land and Sea in Japan* (Nihon sankai meisan zue, first published 1799, hereafter referred to as the *Illustrated Guide*), represented selected animals and plants as assemblages of objects while also highlighting their close connection with local practices of manufacturing goods.[1] This chapter discusses how eclectic modes of representing animals in image and text in this illustrated guide indexed a multitude of ways of expanding knowledge

1 The eclectic nature of materia medica studies (*honzōgaku*) in early modern Japan is pointed out in Federico Marcon, *The Knowledge of Nature and the Nature of Knowledge in Early Modern Japan*. Chicago, IL, and London: Chicago University Press, 2015, p. 255.

Fan Lin and Doreen Mueller (eds), *Picturing Animals and Plants in Early Modern China and Japan. Innovation, Experiments, and Anxieties*. Taylor & Francis Group, 2025
DOI: 10.5117/9789048559091_CH01

and of making it reliable and useful.[2] The *Illustrated Guide* has a preface by Kimura Kenkadō (1736–1802), a successful brewer, rice wine merchant, and industrious amateur scholar who was active in the Kyoto–Osaka region. It is no coincidence that the first volume of five is devoted to the topic of rice wine production. The guide in its entirety was probably put together by several members of Kenkadō's amateur scholar circle in and around Osaka.[3] Its eclectic style shows the intersection of their manifold interests: the empirical study of animals and plants as well as historical and literati pursuits.

The Delicate Matter of Registering Animals in Reference Works

The modalities of representing selected animals in the *Illustrated Guide* advocate certain methods for producing knowledge, and they also demonstrate the importance of adopting a critical attitude towards practices of producing knowledge. The *Illustrated Guide* drew on conventional sources, such as Chinese and Japanese encyclopaedias and manuals of *materia medica*, and it combined them with insights derived from scholarly empirical observation and from local practices. Learned people increasingly valued empirical engagement with the nonhuman environment. They kept animals and plants for systematic observation, and they also dissected and viewed them using scientific tools such as microscopes. By contrast, less educated people, while also engaging directly with the nonhuman environment, did so based on practical knowledge and embodied experience developed in accordance with the requirements of their professional practices. This practical knowledge was, to some extent, empirical, too, and scholars started to take notice of its merits. That said, the key was to be critical in bringing together information derived from diverse sources.

By considering both knowledge derived from canonical sources and from local customs and practices, the *Illustrated Guide* foregrounded the malleability of knowledge: local names for natural phenomena were manifold and often vague, their meanings obscure to those not proficient in the local dialect. How could one be sure whether different local names addressed the same phenomenon in nature? How was one to register local varieties of animals or plants without falling prey

2 The *Nihon sankai meisan zue* has been translated into French by Annick Horiuchi and Daniel Struve, *Guide Illustré des Produits Renommés des Monts et Mers du Japon*. Paris: Collège de France, 2020.

3 Horiuchi and Struve have done extensive research on the authorship of the *Nihon sankai meisan zue*, and they have concluded that it was the product of several authors: Hirase Tessai, the author of the *Nihon Sankai Meibutsu zue* (1754), on which the *Nihon sankai meisan zue* was partially based, Kimura Kenkadō, and the illustrator Shitomi Kangetsu and his successor Nakai Rankō: Horiuchi and Struve, *Guide illustré des produits renommés des monts et mers du Japon*, p. 15.

to the inaccuracies inherent in local names? The *Illustrated Guide* pointed out clashes as well as any overlap between seemingly incompatible sources – reference works such as encyclopaedias and *materia medica*, ancient court poetry (*waka*), and popular verse (*haikai*).

Modes of engaging with animals in reference works, court poetry, and comic verse could not have been more divergent. Court poets of the past were moved deeply by the sights and sounds produced by insects such as bell crickets (*suzumushi*) or waterfowl like the plover (*chidori*).[4] Focusing on a limited number of affecting attributes of animals such as a wistful cry, poets of the past were able to explore their emotional responses to nature. By contrast, popular verse poets thrived on conflating a multitude of seemingly incompatible things to create a titillating and multilayered web of associations. Their attention span was shorter and their view wider as they freely paired the past with the present.[5] They also broke more boundaries, such as the one between the elegant and the vulgar.[6]

Popular reference works like encyclopaedias put animals and plants back into their proper categories, neatly compartmentalising a multitude of objects that one needed to know about.[7] In doing so, encyclopaedias suggested that everything could be objectified into reliable archives of information. At the same time, animals and plants did not simply exist as pieces of visual and textual information. They also generated their own historical and ontological realities as part of the nonhuman environment. The complex entanglements of people and of the nonhuman environment necessitated a constant updating of printed knowledge. For example, extreme weather that contributed to crop failures in the late eighteenth century inspired the compilation of special herbals providing information on how to find, identify, and prepare alternative edible plants in times of famine.[8] Such plants

4 Haruo Shirane, *Japan and the Culture of the Four Seasons: Nature, Literature, and the Arts*. New York: Columbia University Press, 2012, p. 117.

5 Techniques for bringing together things belonging to different categories in image and in text are discussed in Alfred Haft, *Aesthetic Strategies of The Floating World: Mitate, Yatsushi, and Fūryū in Early Modern Japanese Popular Culture*. Leiden and Boston, MA: Brill, 2013.

6 Shirane, *Japan and the Culture of the Four Seasons: Nature, Literature, and the Arts*, pp. 116–119 describes the differences between 'classical' and 'common' birds, for example.

7 Marcon, *The Knowledge of Nature and the Nature of Knowledge in Early Modern Japan*, pp. 77–86 discusses how encyclopaedias tried to present a wealth of information coherently. For more detail on categorising things in encyclopaedias, see Michael Kinski, "Boxes, Fabrics, and Mirrors: On the Contents and the Classification of Popular Encyclopaedias from Early Modern Japan," in Matthias Hayek and Annick Horiuchi eds., *Listen, Copy, Read: Popular Learning in Early Modern Japan*. Leiden and Boston, MA: Brill, 2014, pp. 70–88.

8 For example, Takebe Seian (1712–1782), physician-in-service to Ichinoseki Domain in northeast Japan, wrote the *Minkan Bikōroku* (*Provision of the People in Times of Famine*), first published in 1755. For more information, see Kikuchi Isao, "Ryūminzu ni miru Tōhoku Daikikin," in *Shiroi Kuni no Uta* 10 (2004): 5.

tended to be harder to find and more difficult to prepare and to digest than the staple 'five grains' (*gokoku*) which included rice and millet.

And it was not just the nonhuman environment that was changeable. Language was not a completely reliable repository of knowledge either. In printed reference works, words often had to be supplemented with visual and material evidence to clarify their meanings.[9] The authors of the *Illustrated Guide* were aware of the pitfalls of text and image as they combined excerpts from well-respected sources of knowledge such as the *materia medica* manual *Medicinal Herbs of Japan* (Yamato Honzō, 1709)[10] with observations of local practices and ancient court poetry. The fourth volume focuses on creatures living in the water, especially river fish and frogs. This was a complex field of knowledge. The author of *Medicinal Herbs of Japan*, the scholar Kaibara Ekiken (1630–1714), lamented the lack of information about fish:

> There are many kinds of fish, especially river fish. Every province has different kinds. One cannot possibly cover them all. Some fish exist in some places and not in others. There are also local differences in shape and taste. [...] Few kinds of river and sea fish are mentioned in books on *materia medica*, and not much is explained about them.[11]

Although water creatures were elusive, one could study how local people caught and processed them. This might also yield insights into the creatures themselves. A large part of the *Illustrated Guide* is devoted to descriptions of methods for catching and processing river and sea fish.[12] In doing so, it continued where other reference works stopped: the messy and fragmented realm of local practices. In the spirit of practical learning, Kaibara Ekiken had already pointed out the potential of local practices as sources of knowledge about natural phenomena.[13] The *Illustrated Guide* went beyond this and added further interest by evoking the habitual ways of looking associated with leisurely travellers in Akisato Ritō's *Illustrated Guide to the Capital* (*Miyako Meisho zue*), first published in 1780. Ritō compiled several such guides in the late eighteenth century, all of which were more than simple illustrated travel guides. Providing extensive information pertaining to the history and the

9 The growing importance of detailed illustrations in *honzōgaku* studies to describe things in nature more accurately is noted in Marcon, "All Creatures Great and Small: Tokugawa Japan and Its Animals," p. 32.

10 The *Yamato Honzō* is discussed in detail in Marcon, *The Knowledge of Nature and the Nature of Knowledge in Early Modern Japan*, pp. 87–110.

11 Kaibara Ekiken, *Yamato Honzō*. Kyoto: Nagata Chōbei, 1709, vol. 13, p. 1, preface.

12 Particularly volumes three and four.

13 The entry on *gori* (rockfish) in Kaibara Ekiken, *Yamato Honzō*, vol. 13, p. 6 states that fishermen had various ways of maximising their catch.

cultural significances of local places, they resembled more closely geographic encyclopaedias.[14]

Gaining Knowledge through Leisurely Looking and Encyclopaedic Sampling

In the eighteenth and nineteenth centuries, anything could be studied, even fleeting fashions in the city. Some people collected theatre ephemera and pasted them into scrapbooks.[15] As a result of various people's personal interest, intricate knowledge of popular culture was obtained and processed. Those interested in cursory knowledge could peruse the ready-made compilations of notable things and places in guides compiled by Ritō and others. In the past, specific sites had been notable mainly for their literary associations, but by the eighteenth century, reading about them and viewing illustrations of them entailed the possibility of first-hand experience.[16] As the readers of guides to local places assumed the role of real or would-be travellers delighting in the manifold possibilities of new encounters and insights through leisurely looking and encyclopaedic sampling, they also formed a community of people who shared knowledge about these places.

Gaining knowledge about local places through leisurely looking demanded diverse viewpoints and illustrated travel guides were eager to please as Robert Goree has demonstrated.[17] To satisfy the diverse curiosities of readers, illustrations in guides to local places provided variety by zooming in and out, as we will see.[18] In doing so, they also made readers aware of a wider field of knowledge produced by readers consuming printed guides, encyclopaedias, and other reference works. In this sense, depicting travellers or bystanders looking at local people going about their daily lives became indexical of a larger community of readers and their modes of knowing about local places.

14 For a detailed discussion see Robert Goree. *Printing Landmarks: Popular Geography and Meisho Zue in Late Tokugawa Japan*. Cambridge, MA: Harvard University Asia Center, 2020 and Robert Goree, "*Meisho Zue* and the Mapping of Prosperity in Late Tokugawa Japan," in *Cross-Currents: East Asian History and Culture Review* 23 (June 2017): 73–107.

15 Jonathan Zwicker, "Playbills, Ephemera, and the Historical Imagination in Nineteenth-Century Japan," *The Journal of Japanese Studies* 35/1 (2009): 37.

16 Oka Midori "Nanga: Chinese-Style Landscapes and Literary Poetics," in John Carpenter, *The Poetry of Nature: Edo Paintings from the Fishbein-Bender Collection*. New Haven, CT, and London: Yale University Press, 2019, p. 88.

17 Goree, "*Meisho Zue* and the Mapping of Prosperity in Late Tokugawa Japan," pp. 73–107. Goree, *Printing Landmarks: Popular Geography and Meisho Zue in Late Tokugawa Japan*. Cambridge, MA: Harvard University Asia Center, 2020.

18 Goree, "*Meisho Zue* and the Mapping of Prosperity in Late Tokugawa Japan," p. 83.

Curiosity went beyond merely taking in the local landscape from afar. The space evoked in the illustrations of guides to local places could be wide and expansive, but it also held the promise of sampling things up close for travellers so inclined. A double-page illustration from the *Famous Places of Settsu Province* (Settsu Meisho zue), published between 1796 and 1798, depicts two onlookers spying through a field glass next to an excited group of young people who are gathering *matsutake* mushrooms.[19] The accompanying poem, quoted from the female poet Kaga no Chiyo (1703–1775), hints at the joys of sampling things while travelling:

Takegari ya	Ah, the mushroom hunt!
Aru michi oite	On the road,
Mono no naka	in the thick of things.[20]

The delight taken by these young people in collecting a bumper crop of the elusive *matsutake* mushrooms was also an entertaining visual spectacle for the readers of the *Famous Places of Settsu Province*. Viewing fruitful activities in local places could be as stimulating as acquiring actual local products. The pleasure of swift gratification applied to both. A print in the series titled *Auspicious Pictures of Land and Sea* (*Sankai medetai zue*) by Utagawa Kuniyoshi (1798–1861), which was a humorous take on the titles of printed reference works like the *Illustrated Guide* and the *Illustrated Famous Products of the Mountains and Seas of Japan* (*Nihon Sankai Meibutsu zue*, first published in 1754), compares the burning desire of a fashionable townswoman to see a play in Edo with the enormous efforts of local men pulling a whale ashore with ropes in Hirado Domain in southwest Japan (Figure 1.1).[21] The title of the print, "Can't Wait to See" (*Hayaku mitai*), is a pun on the word "auspicious" (*medetai*) in the series title. Whales were auspicious as they generated prosperity for the local communities who caught and processed them, but they were not easily obtained. Catching a whale was a strenuous and time-consuming undertaking that required the collaboration of many people. By contrast, watching a performance provided swifter and probably more fleeting gratification. Yet, in both cases, auspiciousness was tied to the desire to see or to obtain certain things.

Within the community of real and imaginary travellers, townspeople were keen observers of local things, probably because they were somewhat alienated from

19 Illustrated in Goree, "*Meisho Zue* and the Mapping of Prosperity in Late Tokugawa Japan," p. 90, Figure 6.11.

20 The poem is quoted in Hirase Tessai and Hasegawa Mitsunobu. *Nihon Sankai Meibutsu zue*. Osaka: Shioya Uhei, 1797, vol. 5, p. 58. Poem transcribed and translated by this author.

21 The *Nihon Sankai Meibutsu zue* describes the entire process of the whale hunt, from sighting the whale to bringing it ashore with illustrations: Hirase Tessai and Hasegawa Mitsunobu, *Nihon Sankai Meibutsu zue*, vol. 5 pp. 23–27.

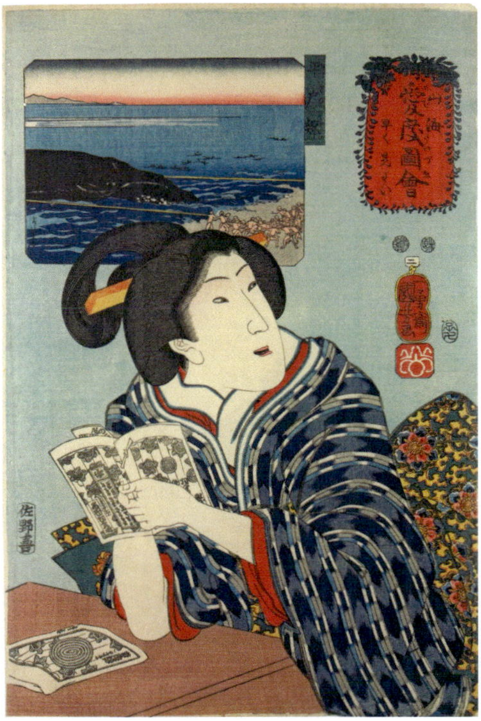

Figure 1.1: Utagawa Kuniyoshi, *Can't Wait To See.* (*Hayaku mitai*), Series: *All the Famous Products of Land and Sea* (Sankai Meisan zukushi), 1852, woodblock print, British Museum Collection.

them. People living in cities were familiar with processed goods, but less so with the sources of these products. Whale products, for instance, circulated throughout the land, but it was near impossible to see a whole whale on land. As Jakobina Arch has demonstrated, whales had to be brought ashore and by then they were usually dead.[22] The rare opportunity to see a whale turned the rotting corpse of one into a spectacle in Osaka in 1823.[23] In this sense, looking at local places for city dwellers also meant connecting in one's mind with the origins of familiar products.

Looking Seriously at the *Illustrated Guide*

The serious and the leisurely were not diametric opposites in the field of popular practice. Both produced knowledge and the illustrations in the *Illustrated Guide* made readers conscious of this. As in illustrated travel guides to famous places, the images in the *Illustrated Guide* tend to depict local practices from some distance,

22 Jakobina K. Arch, *Bringing Whales Ashore: Oceans and the Environment of Early Modern Japan*. Seattle, WA: University of Washington Press, 2018.

23 Arch, *Bringing Whales Ashore: Oceans and the Environment of Early Modern Japan*, pp. 132–139.

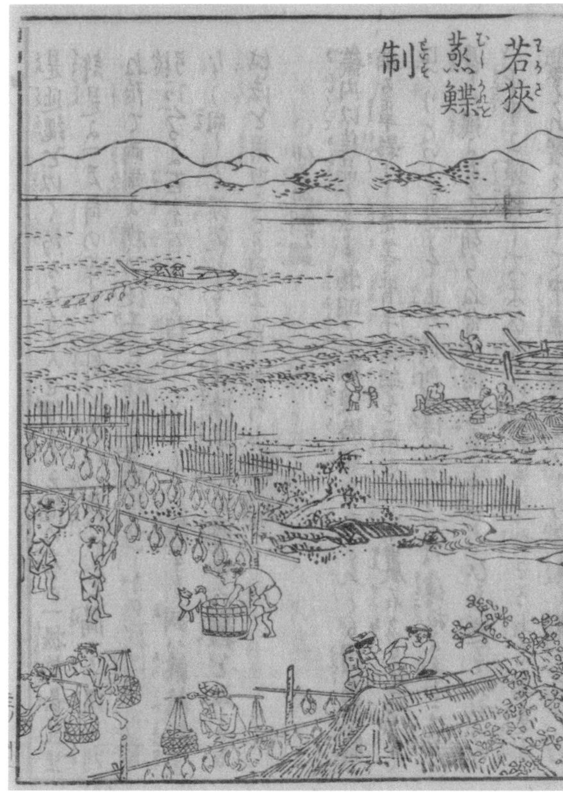

Figure 1.2: Shitomi Kangetsu, Processes for preparing steamed flounder in Wakasa Province, in Hirase Tessai, Kimura Kenkadō, Shitomi Kangetsu, Nakai Rankō. *Illustrated Guide to Famous Products of Land and Sea in Japan* (*Nihon sankai meisan zue*). (Osaka: Yanagiwara Kihei, 1800), Vol. 3, National Diet Library, Tokyo.

situating them in the context of the wider environment. By employing a sweeping view, they evoke the leisurely gaze of the curious traveller. Local people are shown from afar, absorbed in the microcosm of their daily lives, and seemingly unaware of the presence of outsiders (Figure 1.2). This emphasises the gap between the observer and the locals.

At the same time, other kinds of looking are likewise at work in the *Illustrated Guide*. Local women and children, for instance, are depicted looking at fishermen working far out at sea (Figure 1.2). The sea was a precarious place to make a living and depicting locals watching fishermen at work in this way allowed for a more empathetic kind of looking on the part of the readers. Illustrations of this kind also created a sense of a shared community of onlookers. In Figure 1.3, for example, uninvolved bystanders are pointing at fishing boats in the distance, probably discussing what they are seeing. The children of local fishermen also look and point, but their focus is on the men and women processing bonito close-by on the shore. The two different viewpoints co-exist in this scene, demonstrating that looking had become a serious matter as its modalities indexed different levels of engagement with local practices.

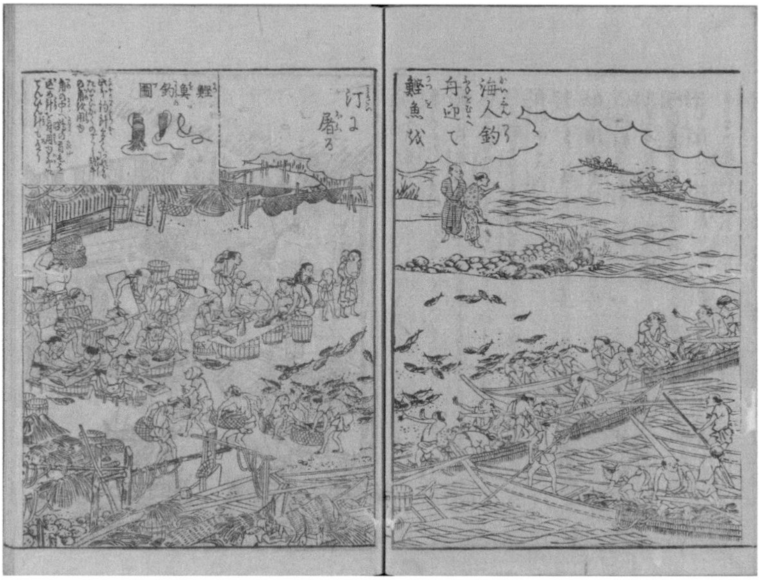

Figure 1.3: Shitomi Kangetsu, Processes for processing bonito by the shore, in Hirase Tessai, Kimura Kenkadō, Shitomi Kangetsu, Nakai Rankō. *Illustrated Guide to Famous Products of Land and Sea in Japan (Nihon sankai meisan zue).* (Osaka: Yanagiwara Kihei, 1800), Vol. 4, National Diet Library, Tokyo.

This was a departure from the illustrations in the *Illustrated Famous Products of the Mountains and Seas of Japan* of roughly thirty years earlier. Here, looking at something in the distance was mainly done by parties directly involved in obtaining desirable objects, such as whales. Without distractions in the form of the gazes of curious bystanders, the illustrations focus the reader's attention directly on the local fishermen, who, although dwarfed by the menacing presence of a giant whale endowed with huge eyes and emitting a fountain of water, manage to catch it and successfully bring it ashore.[24] This depiction furnishes the spectacle of the whale hunt with a sense of danger and self-congratulatory achievement. Presenting this spectacle through the eyes of uninvolved bystanders would have detracted from the enormous efforts made by local fishermen. In this way, the *Illustrated Famous Products of the Mountains and Seas of Japan* foregrounded how whales were obtained and processed by the locals.

By contrast, scholars of materia medica (*honzōgaku*) sought visual accuracy in representing animals and plants as this allowed them to train their observational skills and to differentiate between similar specimens.[25]They required such specialised skills as they often enjoyed privileged access to collections of specimens. This was probably not the case for most readers of the *Illustrated Famous Products*

24 Illustration in Hirase Tessai and Hasegawa Mitsunobu, *Nihon Sankai Meibutsu zue*, vol. 5, p. 24.

25 Federico Marcon, "All Creatures Great and Small: Tokugawa Japan and Its Animals," in Robert T. Singer and Kawai Masatomo eds., *The Life of Animals in Japanese Art*, Princeton, NJ: Princeton University Press, 2019, p. 32.

of the Mountains and Seas of Japan. Although its illustrations were not intended to train readers in close observational skills, the book still raised awareness of the combined power of looking and of consulting existing sources to produce knowledge about animals and plants. For example, an illustration depicting mischievous water spirits called 'river boys' (*kawatarō*) wrestling at a river's edge highlights their furry bodies, playful nature, and the fact that they could be encountered near water.[26] The depicted place could have been anywhere. The text is more informative and tells us that there were many river boys in the Kantō region, where they were called river children (*kawa warawa*).[27] In addition, Kaibara Ekiken's *Medicinal Herbs of Japan* is quoted, warning that these water spirits may drag people into the water, causing a madness that could be cured with star anise.[28]

The images in the *Illustrated Guide* further developed the awareness of the importance of careful observation and accurate visual representation. In this book, natural phenomena were no longer merely curiosities for the casual onlooker seeking novelty in illustrated travel guides to famous places, or sources of prosperity for local people as depicted in the *Illustrated Famous Products of the Mountains and Seas of Japan*. Both viewpoints could be combined, stimulating a contemplation of different levels of engagement with things in nature, and the knowledge that could be derived from these. An image depicting a young samurai observing local fishermen illustrates looking for the purpose of learning (Figure 1.4). He stands beside his elderly mentor who is pointing at fishermen catching small and slippery rockfish in the Kamo River. The young samurai represents an elite that valued observing local practices in the spirit of practical learning.

Some illustrations are reminiscent of the scientific gaze of natural studies, drawing readers' attention to animals and plants in isolation and in close-up detail.[29] In the top left corner of the scene depicting local fishermen and women processing bonito by the seashore, various hooks for catching fish are assembled in a rectangular cartouche (Figure 1.3). A cloud-shaped cartouche shows bees in different stages of development above a scene depicting honey-making in Kumano.[30] Just below the cartouche, a playful dog pursues a swarm of bees on

26 Illustration in Hirase Tessai and Hasegawa Mitsunobu, *Nihon Sankai Meibutsu zue*, vol. 3, p. 25. For a discussion of the diverse cultural significances of *kappa*, see Michal Dylon Foster "The Metamorphosis of the Kappa: Transformation of Folklore to Folklorism in Japan." *Asian Folklore Studies*, 57/1 (1998): 1–24.

27 Hirase Tessai and Hasegawa Mitsunobu, *Nihon Sankai Meibutsu zue*, vol. 3, p. 25.

28 Hirase Tessai and Hasegawa Mitsunobu, *Nihon Sankai Meibutsu zue*, vol. 3, p. 25.; Kaibara Ekiken, *Yamato Honzō*, vol. 16, p. 21.

29 The role of faithful reproductions of naturalia through techniques such as ink rubbing in *honzōgaku* studies is discussed in Maki Fukuoka, *The Premise of Fidelity: Science, Visuality, and Representing the Real in Nineteenth-Century Japan*. Stanford, CA: Stanford University Press, 2013.

30 Illustration in Hirase Tessai, Kimura Kenkadō, Shitomi Kangetsu, and Nakai Rankō. *Nihon sankai meisan zue*, Osaka: Yanagiwara Kihei, 1800, vol. 2, p. 17.

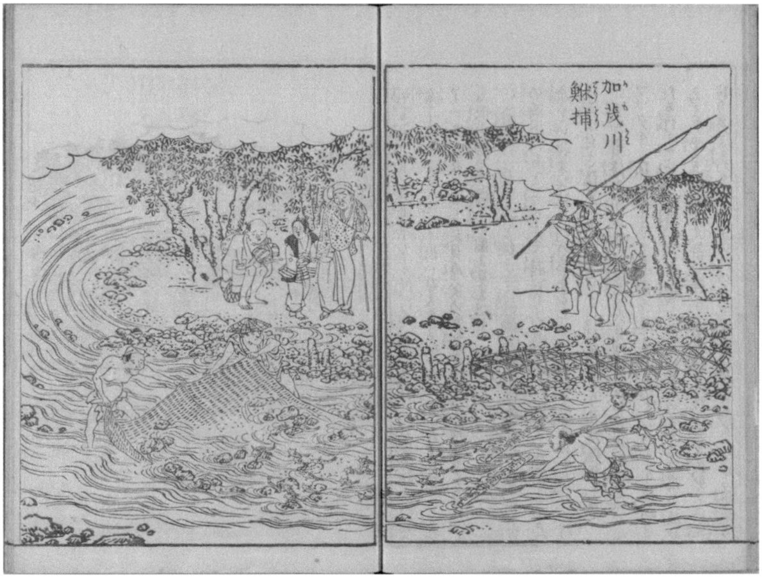

Figure 1.4: Shitomi Kangetsu, Catching gori in Kamo River, in Hirase Tessai, Kimura Kenkadō, Shitomi Kangetsu, Nakai Rankō. *Illustrated Guide to Famous Products of Land and Sea in Japan* (*Nihon sankai meisan zue*). (Osaka: Yanagiwara Kihei, 1800), Vol. 4, National Diet Library, Tokyo.

the roof of a beekeeper's house. Readers of illustrated books, including illustrated travel guides to famous places, were familiar with cartouches providing further information, but cartouches highlighting things and assembling them for the purpose of comparing their forms were new. This type of illustration unites two seemingly clashing registers of knowledge: comparing the forms of things in the cartouche and contextualising these things as part of local practices in the scene below.

Examining Critically: Singing Fish versus Singing Frogs

There was value in an eclectic approach to looking at animals and plants. It afforded a glance at matters from different perspectives, making readers aware of the artifice inherent in representing things for the purpose of studying them. Even deeper insights could be produced by collecting actual specimens. Collections allowed their owners to directly compare things and go beyond their visual appearance. However, introducing a specimen into a collection meant taking it out of its original context. In addition, items often had to be preserved to be made collectible. Kimura Kenkadō's collection of seashells drew attention to the manifold forms of shells, but they were no longer alive nor in their natural surroundings.[31] The shell collection

31 The shell collection is illustrated in Osaka Museum of History ed., *Kimura Kenkadō: Naniwa chi no kyojin Tokubetsuten botsugo 200-nen* (Kyoto: Shibunkaku shuppan, 2003).

possibly inspired Itō Jakuchū's painting of shells by the seashore.[32] Jakuchū carefully captured the forms of the shells, but their assemblage on the painted seashore was artificial. Preserving the integrity of natural objects required placing them back into their original context.

All methods for studying things were fallible. Just like objects assembled in collections, things reproduced in print could be manipulated, albeit through image and text. Considering these seemingly endless possibilities, the *Illustrated Guide* advocated the need for a critical attitude towards examining evidence. This is especially the case in a section of the fourth volume, which discusses rockfish, a local delicacy in various regions of Japan. Rockfish were tricky to capture, in several respects. Their forms and local names were manifold. Being small and slippery, they eluded the human eye even in clear water. Local fishermen found ways to catch them by exploiting their natural behaviour. When in danger, the fish attach themselves to rocks with a sticky fin on their belly, in which case they can be scooped up with special tools, which varied from region to region.

Kaibara Ekiken had already noted in the *Medicinal Herbs of Japan* that local fishermen had various ways of catching rockfish.[33] In the spirit of practical learning, he also pointed out that rockfish were colloquially called *gori* or *kajika*, the latter literally meaning 'river deer', a vague name that could denote frogs as well as fish.[34] The *Illustrated Guide* describes local differences in naming and catching the fish in detail. The first such description coincides with the image of the young samurai observing local fishermen catching rockfish in Kamo River (Figure 1.4):

> Two straw mats are combined and submerged. Rocks are piled onto the mat. Two people lift one side while another person dredges the bottom of the river [...]. The fish come up and are caught as they try to hide by attaching themselves to the rocks on the mat. Fish and rocks are then scooped up together. This is called 'shoving gori' (*gorioshi*).[35]

The *Illustrated Guide* did not content itself with merely adding to observations already made in the *Medicinal Herbs of Japan*. It had bigger problems to address. Local people tended to confuse different things that dwelled in the same places under the

32 Jakuchū painted this between 1761 and 1765 as part of a Buddhist devotional project of a series of silk hanging scrolls depicting the realm of sentient beings. Illustrated and discussed in Yukio Lippit. *Colorful Realm: Japanese Bird-and-flower Paintings by Itō Jakuchū* (Chicago, IL: University of Chicago Press. 2012), p. 118.

33 Kaibara Ekiken, *Yamato Honzō*, vol. 13, p. 6.

34 Kaibara, *Yamato Honzō*, vol. 13, p. 6.

35 Hirase Tessai, Kimura Kenkadō, Shitomi Kangetsu, and Nakai Rankō, *Nihon sankai meisan zue*, vol. 4, p. 19.

same name. According to the *Illustrated Guide*, local people called anything that made a noise in the water *kajika*.[36] This even included frogs singing with melodious voices. In ancient poetry they had a special name, *kawazu*, but this had largely been forgotten. The dwelling places of singing frogs, clear mountain streams, also happened to be the habitat of rockfish. The voices of rockfish, it was said, resembled a crunching noise or the high-pitched ping that came from the bottom of a porcelain bowl when struck. Kaibara Ekiken found this ping charming.[37]

The popular practice of subsuming different creatures that shared the same habitat under one name was not conducive to the careful documentation of one's observations. In addition, the word *kajika* itself proved a real bone of contention. It was as thoroughly colloquial as other local names such as *gori*, but some people pretended otherwise. The people in question, according to the *Illustrated Guide*, were popular verse (*haikai*) poets.[38] Ever since the Kan'ei Era (1624–1644), they had sacrificed accuracy for the sake of creating witty poetry.[39]

Poets of the recent past were fond of the word *kajika* to denote various things that made noises in the water. The revered *haikai* poet Matsuo Bashō (1644–1694) had used the proper poetic word for singing frogs, *kawazu*, but in his oft-cited *Old Pond* (*Furu ike*) poem the frog is notable simply for plopping into the water, not for its beautiful voice, which had been celebrated in ancient court poetry.[40] The pursuit of peripheral and spectacular effects, such as plops, had turned poets' attention away from carefully registering the actual creature. Their disingenuity went even further than this. Various poems using the word *kajika* were attributed to the monk-poet Saigyō (1118–1190), celebrated for displaying some measure of eccentricity and a love of cherry blossoms. According to the *Illustrated Guide*, Saigyō was probably not the author of these poems because they could not be traced back to historical poetry collections.[41]

In sum, the *Illustrated Guide* contended that moulding the past as a mirror image of the present was not helpful for carefully distinguishing things. Erasing the gap between the past and the present, the high and the low, and playing with words and their meanings was an excusable offence for popular verse poets. The spectacular

36 Hirase Tessai, Kimura Kenkadō, Shitomi Kangetsu, and Nakai Rankō, *Nihon sankai meisan zue*, vol. 4, p. 19.

37 Kaibara Ekiken, *Yamato Honzō*, vol. 13, p. 6.

38 Hirase Tessai, Kimura Kenkadō, Shitomi Kangetsu, and Nakai Rankō, *Nihon sankai meisan zue*, vol. 4, p. 32.

39 Hirase Tessai, Kimura Kenkadō, Shitomi Kangetsu, and Nakai Rankō, *Nihon sankai meisan zue*, vol. 4, p. 32.

40 "The Old Pond/A Frog Jumps in/The Sound of Water," translated in Peter France (ed.), *The Oxford Guide to Literature in English Translation*. Oxford: Oxford University Press, 2000, p. 239.

41 Hirase Tessai, Kimura Kenkadō, Shitomi Kangetsu, and Nakai Rankō, *Nihon sankai meisan zue*, vol. 4, pp. 32–33.

effects of word play allowed them to show off their literary skills, but this was not acceptable for scholars seeking to order things in nature with precision. Here, the *Illustrated Guide* even caught out the venerable Master Kaibara, who had argued in his *Medicinal Herbs of Japan* that the name *kajika* appeared in ancient poems.[42] This, the *Illustrated Guide* stated, was a mistake.[43] The indiscriminate use of the name *kajika* to denote things that made noises in the water symbolised the failure of the present to register things properly, and the *Illustrated Guide* sought to prove this by carefully tracing the history of the word.

In contrast to poets of the present, it stated, poets of the past had listened carefully, singling out singing frogs by their proper name, *kawazu*, which left no doubt as to what it denoted. The *Illustrated Guide* quotes various poems from historical poetry anthologies, such as the *Collection of Ten Thousand Leaves* (Manyōshū, c. 759 CE), to demonstrate how poets of the past had registered the voice of singing frogs, including the times when the frogs sang, in summer and in autumn, twice a day, at noon and in the deep of night. Although these poets had been mainly interested in singing frogs for poetic reasons, they had nevertheless closely observed their distinguishing attributes.

One of the reasons why singing frogs were given their proper name in ancient poems was the fact that court poets were only concerned with attractive creatures singing in pleasant environments like clean mountain streams. They did not bother with common frogs squawking in muddy rice paddies and filthy swamps. Over time, poets became less discerning, and they started to situate their poetry in lesser places, using colloquial names for things. It was during the lifetime of Kamo no Chōmei (1155–1216) that the waters got muddied even in court poetry.[44] Poets started to mix up *kawazu* and the colloquial *kajika*, and by the time the *Illustrated Guide* was published, most people had forgotten the old name *kawazu*, which had always specifically denoted singing frogs endowed with a beautiful voice dwelling in clear mountain streams.[45]

Popular verse poets argued that using colloquial names made their poems modern, enabling everyone to understand what they were saying.[46] However, in the case of *kajika*, they sacrificed accuracy for popularity. People's memories were

42 Kaibara Ekiken, *Yamato Honzō*, vol. 13, p. 7.

43 Hirase Tessai, Kimura Kenkadō, Shitomi Kangetsu, and Nakai Rankō, *Nihon sankai meisan zue*, vol. 4, p. 32.

44 Hirase Tessai, Kimura Kenkadō, Shitomi Kangetsu, and Nakai Rankō, *Nihon sankai meisan zue*, vol. 4, p. 34.

45 Hirase Tessai, Kimura Kenkadō, Shitomi Kangetsu, and Nakai Rankō, *Nihon sankai meisan zue*, vol. 4, p. 37.

46 Haft, *Aesthetic Strategies of The Floating World: Mitate, Yatsushi, and Fūryū in Early Modern Japanese Popular Culture*, p. 70.

fickle, and the ability of words to capture things in nature was unreliable. Misuse of terms for the sake of effect had a long history and had become entrenched in local practices for registering things in nature. Kaibara Ekiken had argued as much in the *Medicinal Herbs of Japan*, noting that many colloquial fish names had been passed down wrongly, and people had accepted the erroneous names without question.[47] Accepting something as the truth purely based on custom was an error of judgment.

Although Kaibara Ekiken had indicated this, he had ultimately not acted upon it. The *Illustrated Guide* did. It remarks that Kamo no Chōmei had described singing frogs in his *Nameless Notes* (Mumyosho, ca. 1211–1216), pointing out that these frogs lived in the mountain streams of Ide, their colour was dark, and they were small.[48] They also lived mostly in the water, unlike common frogs, which only spend their early life in the water. In addition, the croaking of these frogs in the nighttime moved people's hearts. The *Illustrated Guide* added more information about the local varieties of singing frogs. *Kashima* frogs have a mottled pattern; they are very fast and hard to catch. At night, their voices resemble the robin's call, and in the summer months they sing once every hour. At noon, their call sounds like that of a shrike.[49]

The *Illustrated Guide* admits that, without due care, it was easy to confuse singing frogs and singing fish. They shared certain characteristics: they were small, fast, and not easily caught, and dwelled in mountain streams from where their high-pitched voices rose at night. Nevertheless, they were different things. Recognising them as such required exposing the name *kajika* as colloquial and vague. And when words failed to be precise, they needed to be backed up with images. Consequently, the *kajika* section closes with an illustration highlighting the outstanding physical features of singing frogs, particularly their unwebbed fingers with rounded fingertips (Figure 1.5). The fingers of the frogs jump out at the viewer, being shown both front and back. The caption clears up the confusion about the name *kajika*: "As already explained, [singing frogs] have a beautiful appearance and voice. *Kajika* is a colloquial name. In court poems, the name is not pronounced *kajika* but *kawazu*. This is proven in *The Collection of Ten Thousand Leaves*."[50]

In highlighting the unwebbed fingers of singing frogs, the illustration brought home to readers the importance of carefully observing small details to correctly

47 Kaibara Ekiken, *Yamato Honzō*, vol. 13, p. 1.

48 Hirase Tessai, Kimura Kenkadō, Shitomi Kangetsu, and Nakai Rankō, *Nihon sankai meisan zue*, vol. 4, p. 34.

49 Hirase Tessai, Kimura Kenkadō, Shitomi Kangetsu, and Nakai Rankō, *Nihon sankai meisan zue*, vol. 4, p. 33.

50 Hirase Tessai, Kimura Kenkadō, Shitomi Kangetsu, and Nakai Rankō, *Nihon sankai meisan zue*, vol. 4, p. 37.

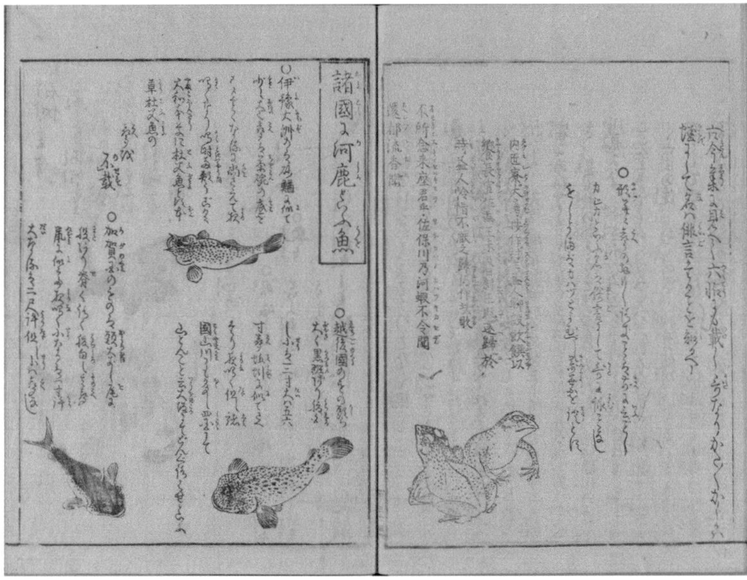

Figure 1.5: Shitomi Kangetsu, Singing frogs and fish called kajika in various regions in Japan, in Hirase Tessai, Kimura Kenkadō, Shitomi Kangetsu, Nakai Rankō. *Illustrated Guide to Famous Products of Land and Sea in Japan (Nihon sankai meisan zue).* (Osaka: Yanagiwara Kihei, 1800), Vol. 4, National Diet Library, Tokyo.

identify an object. Although the illustration is not carefully detailed otherwise, it provided an emphatic statement in favour of the value of careful observation, thereby feeding into the wider discourse about visual accuracy in the field of natural studies in eighteenth-century Japan. Furthermore, within this field, the dissecting gaze of the naturalist closely observing and registering the appearance of things in nature did not necessarily exclude the historical meanings of this thing in the cultural imagination. Court poetry had long produced a historical trajectory of secondary nature that impacted on how animals and plants were perceived and registered.[51]

At the same time, this cultural history of animals and plants was considered with a novel critical fervour. While the *Illustrated Guide* claimed that ancient poets had registered things in nature with some care, it also indicated that they had prioritised aesthetic concerns in doing so. They only registered beautiful things that drew attention to meaningful times or places, giving rise to a multitude of poetic associations. If one was to observe nature more comprehensively, one needed to look at the common creatures dwelling in less savoury places, too. Capturing the diversity of things meant overcoming the cultural preference for beautiful things. In addition, popular verse poets did have a point in arguing in favour of colloquial names, but their pursuit of the spectacular led to carelessness.

51 Shirane, *Japan and the Culture of the Four Seasons*. The futility of the separation of nature and culture is discussed in Bruno Latour, *Politics of Nature: How to Bring the Sciences into Democracy* (Cambridge, MA: Harvard University Press, 2004).

The Value of Assembling Critically

Scholars of practical learning, such as Kaibara Ekiken, also took note of local practices for registering things in nature. In addition, they were eclectic in combining elite and common sources of knowledge. Unlike popular verse poets, however, they pointed out the origins of these sources. In *Medicinal Herbs of Japan*, Kaibara referenced Chinese and Japanese works such as the *Compendium of Materia Medica* (Bencao Gangmu, first completed in 1578) and *Enlightening Illustrations* (Kinmō zui, first published in 1666), and combined these with local knowledge. This enabled him to link the Chinese name for rockfish, *tofugyo* (Chinese: *dufu yu*), taken from the *Compendium of Materia Medica*, to the colloquial *gori* and *kajika*.[52] However, Kaibara did go too far in arguing that the name *kajika* went back to ancient court poetry. The *Illustrated Guide* proved otherwise by tracing the history of the term.

The *Illustrated Guide* applied a critical historical consciousness, and it expanded on the local names listed in the *Medicinal Herbs of Japan* by delving more deeply into their origins and local meanings.[53] This revealed that local names sometimes hinted at the conspicuous behaviour of certain creatures. Some rockfish bury themselves in the mud to hunt for prey. Locals therefore called them *chichikaburi* – *chichi* meaning 'earth' and *kaburi* 'to cover.'[54] Singing frogs displayed similar behaviour in the summer months, making this the only time when they could be caught more easily. Other rockfish only existed in certain places. *Arare-uo*, literally 'hail fish,' were only found in Echizen Province, and were called so because they had the curious habit of floating belly-up in inclement weather.[55]

Local varieties of rockfish and their names were seemingly endless. Delving into the hodgepodge of local lore revealed an apparently inextricable tangle of diversity, which could only be made comprehensible by deconstructing it into its constituent parts and organising it through observation and representation. To this effect, the *Illustrated Guide* presents an assemblage of illustrations of various rockfish, all of which were called *kajika* locally (Figures 1.5, 1.6). The illustrations have added explanations, and, as a whole, the assemblage demonstrates a correct order for things in nature. Place came first, as it determined variations in the forms and the names of things. Next came comparisons with similar creatures, and secondary information derived from reference works.

52 Kaibara Ekiken, *Yamato Honzō*, vol. 13 p. 7.

53 Hirase Tessai, Kimura Kenkadō, Shitomi Kangetsu, and Nakai Rankō, *Nihon sankai meisan zue*, vol. 4, p. 32.

54 Hirase Tessai, Kimura Kenkadō, Shitomi Kangetsu, and Nakai Rankō, *Nihon sankai meisan zue*, vol. 4, p. 32.

55 Hirase Tessai, Kimura Kenkadō, Shitomi Kangetsu, and Nakai Rankō, *Nihon sankai meisan zue*, vol. 4, p. 32.

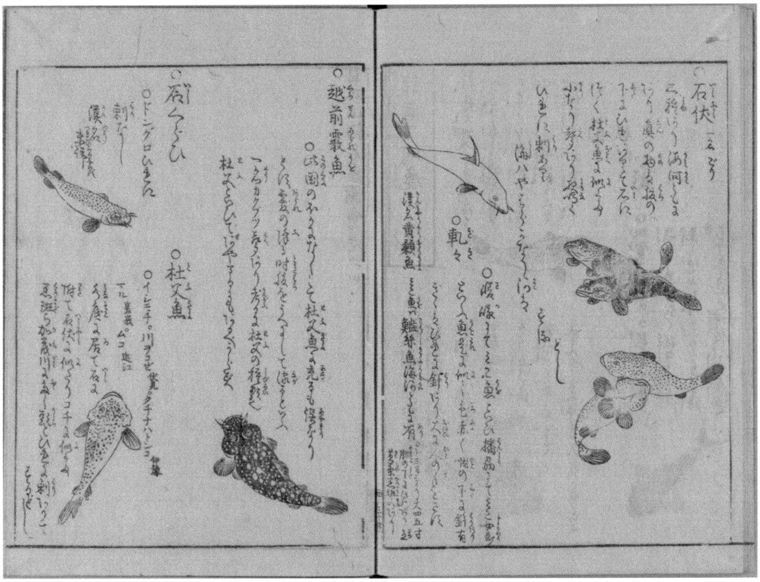

Figure 1.6: Shitomi Kangetsu, Singing frogs and fish called kajika in various regions in Japan, in Hirase Tessai, Kimura Kenkadō, Shitomi Kangetsu, Nakai Rankō. *Illustrated Guide to Famous Products of Land and Sea in Japan* (*Nihon sankai meisan zue*). (Osaka: Yanagiwara Kihei, 1800), Vol. 4, National Diet Library, Tokyo.

Rockfish look alike at first glance, but close examination reveals differences in the size and shape of heads and fins, markings on the body, and the absence or presence of spikes. As has been mentioned earlier, rockfish have a fin on their belly with which they attach themselves to rocks. There is a marked difference between the texture of the belly fins of sea-dwelling and river-dwelling types – the belly fins of rockfish in rivers have sharp spikes, whereas the spikes of sea-dwelling types are soft.[56] The illustrations (Figure 1.6) highlight this difference by displaying the dorsal and the ventral views of both types. In addition, some of the fish that were called *kajika* by the locals were not rockfish at all. A catfish called *gigi* is the odd-one-out in the assemblage (Figure 1.6). Among the stocky and mottled bodies of rockfish, its slender form and monochrome colour stand out. In fact, the text notes that its body is red. In the absence of colour, the illustrations highlight this difference by leaving its body blank.

The illustrations thus emphasise small differences in the appearance of rockfish, impressing on readers the importance of careful observation and comparison. The added explanations helped readers create a mental map by connecting the varying forms to local places. Rockfish in Echigo Province have a large head and a black mottled pattern whereas the backs of rockfish in the domain of Kaga are completely black (Figure 1.5). Rockfish in Iyo Province resemble loaches in the

56 Hirase Tessai, Kimura Kenkadō, Shitomi Kangetsu, and Nakai Rankō, *Nihon sankai meisan zue*, vol. 4, p. 37.

sand (Figure 1.5).[57] The *Illustrated Guide* also did away with superfluous details. In the *Enlightening Illustrations* encyclopaedia that had been published roughly a century earlier, rockfish appear in the water and surrounded by rocks, expressing their rightful place as water creatures in the traditional East Asian cosmology of heaven, earth, and the human world.[58]

However, this manner of representation detracted from the visual appearance of the fish. The text that accompanied the illustrations in the *Illustrated Guide* provided information as to where exactly different kinds of rockfish dwelled. Certain types of information, for instance on habitat, were expressed more accurately through words, and illustrations worked better for highlighting differences in patterns, body shapes, and size. In addition, pinning down the place of things in nature required more information, and transcending the limits of local lore and direct observation. This information relied on secondary channels of information such as the encyclopaedia *Categorical Miscellany of Yamato Names* (Wamyō ruijushō, circa 934), compiled by Minamoto no Shitagō (911–983).[59]

Based on an entry in this encyclopaedia, Kaibara Ekiken noted that the *Food Classic* (*Shijing*) by Cui Hao (active 408–452) recorded that *ishibushi* rockfish hide among rocks at the bottom of rivers, hence the literal meaning of their name – 'rock-dweller.'[60] The *Illustrated Guide* added that the *Categorical Miscellany of Yamato Names* did not mention *gori*, only *ishibushi*.[61] *Gori* was too recent and too colloquial a name to have made it into this classic encyclopaedia. As a result of his study of local names for rockfish, Kaibara Ekiken was able to note that the colloquial term *gori* and the more dignified word *ishibushi* denoted the same thing. Bringing together information from diverse sources enabled the making of new connections, but this required a critical mind and the ability to judge the reliability of sources.

The *Illustrated Guide* impressed this on its readers by pointing out routes of knowledge to illustrations of rockfish. Local lore had it that rockfish called in the night, but their voices were not always the same. In Iyo Province, rockfish sounded like earthworms, whereas in Kaga Domain their calls resembled the squeaking of rats.[62] In Echigo Province, their calls evoked the melodious ping emitted from the

57 Hirase Tessai, Kimura Kenkadō, Shitomi Kangetsu, and Nakai Rankō, *Nihon sankai meisan zue*, vol. 4, p. 37.

58 Illustration in Nakamura Tekisai ed., *Kinmō Zui* (Place unknown: Yamagataya, 1666), vol. 11, p. 16.

59 The Wamyō ruijushō is discussed in Steininger, Brian. *Chinese Literary Forms in Heian Japan: Poetics and Practice*. Boston, MA: Harvard University Asia Center, 2017.

60 Kaibara Ekiken, *Yamato Honzō*, vol. 13, p. 6.

61 Hirase Tessai, Kimura Kenkadō, Shitomi Kangetsu, and Nakai Rankō, *Nihon sankai meisan zue*, vol. 4, p. 32.

62 Hirase Tessai, Kimura Kenkadō, Shitomi Kangetsu, and Nakai Rankō, *Nihon sankai meisan zue*, vol. 4, p. 37.

bottom of a porcelain bowl when struck.[63] The explanations note that nothing was mentioned about such voices in the *Medicinal Herbs of Japan*.[64] Kaibara Ekiken had mentioned in the *Medicinal Herbs of Japan* that the clear and high-pitched cry of *gori* rockfish was adorable when heard at night.[65]

Collecting information from various sources showed up knowledge gaps and discrepancies. Resolving these required a discerning mind. The widespread availability of printed reference works created the possibility of producing informed readers.[66] Readers of the *Illustrated Guide* could be expected to have recourse to the same Chinese and Japanese reference works that informed manuals of *materia medica* like the *Medicinal Herbs of Japan* and reference works like the *Illustrated Guide*. They may not have had the education of scholars like Kaibara Ekiken, but they did have access to the same sources. With this in mind, the *Illustrated Guide* presented information in such a way that it made readers aware of the possibility of cross-referencing particulars from various sources and in this way increase their knowledge.

Conclusion

In combining the modalities of leisurely and empirical looking with modes of encyclopaedic sampling, the *Illustrated Guide* followed epistemic practices that were current in printed reference works, such as illustrated travel guides to famous places. Like illustrated travel guides to famous places, its illustrations hinted at a rising level of self-consciousness regarding the relative place of readers who tapped into knowledge conveyed through printed media. This place was a matter of constant negotiation vis-a-vis local places and practices as well as the imagined community of readers of reference works. The growing number and variety of reference works, which included both printed media and manuscripts, heightened a sense of urgency to advocate critical scrutiny, especially as information derived from increasingly diverse sources showed up incongruities that had to be resolved or at least thematised. The eclectic combination of leisurely, empirical, and encyclopaedic modes of knowing thus caused some measure of anxiety about the authenticity of knowledge and the usefulness of practices of producing knowledge.

It can be argued that, in response to these developments, the *Illustrated Guide* acted as an intervention by highlighting the value of cultivating self-consciousness

63 Hirase Tessai, Kimura Kenkadō, Shitomi Kangetsu, and Nakai Rankō, *Nihon sankai meisan zue*, vol. 4, p. 37.

64 Hirase Tessai, Kimura Kenkadō, Shitomi Kangetsu, and Nakai Rankō, *Nihon sankai meisan zue*, vol. 4, p. 37.

65 Kaibara Ekiken, *Yamato Honzō*, vol. 13, p. 6.

66 For more information, see Yokota Fukuhiko, *Dokusho to Dokusha* (Tokyo: Heibonsha, 2015).

about the position of readers relative to empirical realities and to the growing body of knowledge. Merely assembling knowledge no longer sufficed; to hold true, it had to be cross-referenced from a position of historical scrutiny. As the discussion of the naming of rockfish and frogs has demonstrated, this also came from the awareness that the eye and the mind could be fooled into morphing one thing into another, and that things could be reshaped and repurposed through physical manipulation. Cultural practices of naming and representing things changed over time, too. In response, the *Illustrated Guide* advocated vigilance. Knowledge needed to be updated constantly. In showing different ways to register things in nature, the *Illustrated Guide* highlighted the power of image and text to shape knowledge. The *kajika* section of the *Illustrated Guide* reads like an admonishment, reminding readers of their duty to examine evidence critically. Bringing in the field of popular practice could reveal unknown connections, such as between local rockfish and Chinese *tofugyo* (*dufu yu*), but the production of useful insights required weeding out inconsistencies and make-believe. It was always best to pin down the thing to its source – both in nature and in available sources.

Glossary

Akisato Ritō	秋里籬嶋
Arare-uo	霰魚
Chichikaburi	鯒
Bencao Gangmu (J: *Honzō Kōmoku*)	本草綱目
Chidori	千鳥
Cui Hao	崔浩
Echigo	越後
Furu ike	古池
Gigi	鮠
Gokoku	五穀
Gori	鮴
Gorioshi	ごり押し
Haikai	俳諧
Hasegawa Mitsunobu	長谷川光信
Hirase Tessai	平瀬徹斎
Honzōgaku	本草学
Ishibushi	石伏
Itō Jakuchū	伊藤若冲
Iyo Province	伊予国
Kaga Domain	加賀国

Kaga no Chiyo	加賀の千代
Kaibara Ekiken	貝原益軒
Kajika	河鹿
Kamo no Chōmei	鴨長明
Kamo River	鴨川
Kantō	関東
Kawatarō	川太郎
Kawa warawa	河童
Kawazu	蛙
Kimura Kenkadō	木村蒹葭堂
Kinmō zui	訓蒙圖彙
Kumano	熊野
Manyōshū	万葉集
Matsuo Bashō	松尾芭蕉
Matsutake	松茸
Miyako Meisho zue	都名所図会
Minamoto no Shitagō	源順
Mumyosho	無名抄
Nakamura Tekisai	中村惕斎
Nihon sankai meibutsu zue	日本山海名物図会
Nihon sankai meisan zue	日本山海名産図絵
Saigyō	西行
Settsu meisho zue	摂津名所図会
Shijing	食經
Shitomi Kangetsu	蔀関月
Suzumushi	鈴虫
Tofugyo (C. Dufu yu)	杜父魚
Waka	和歌
Wamyō ruijushō	和名類聚抄
Yamato honzō	大和本草

Bibliography

Primary Sources

Akisato Ritō秋里籬島 ed., Takehara Shunchōsai 竹原春朝斎 (illustrations). *Settsu Meisho zue*. Osaka: Morimoto Tarō, 1798. Collection of National Diet Library Japan.

Hirase Tessai 平瀬徹斎 (author), Hasegawa Mitsunobu 長谷川光信 (illustrations). *Nihon Sankai Meibutsu zue* 日本山海名物図会. Osaka: Shioya Uhei, 1797. Collection of National Diet Library Japan.

Hirase Tessai 平瀬徹斎, Kimura Kenkadō 木村蒹葭堂, Shitomi Kangetsu 法橋關月and Nakai Rankō 中井藍江. *Nihon sankai meisan zue* 日本山海名産図会. Osaka: Yanagiwara Kihei, 1800. Collection of National Diet Library Japan.

Kaibara Ekiken 貝原益軒. *Yamato Honzō* 大和本草. Kyoto: Nagata Chōbei, 1709, Collection of National Diet Library Japan.

Nakamura Tekisai 中村惕斎ed. *Kinmō Zui* 訓蒙図彙. Place unknown: Yamagataya, 1666.

Secondary Sources

Arch, Jakobina K. *Bringing Whales Ashore: Oceans and the Environment of Early Modern Japan*. Seattle, WA: University of Washington Press, 2018.

Clark, Timothy ed. *Hokusai: Beyond the Great Wave. London: The British Museum*, 2017.

Cwiertka, Katarzyna. "Meibutsu and Commercialized Travel in Early Modern Japan," in *Branding Japanese Food: from Meibutsu to Washoku.* Honolulu: University of Hawaii Press, 2020, pp. 49-74.

Foster, Michal Dylon. "The Metamorphosis of the Kappa: Transformation of Folklore to Folklorism in Japan," *Asian Folklore Studies* 57/1 (1998): 1–24.

France, Peter ed. *The Oxford Guide to Literature in English Translation.* Oxford: Oxford University Press, 2000.

Fukuoka, Maki. *The Premise of Fidelity: Science, Visuality, and Representing the Real in Nineteenth-Century Japan*. Stanford, CA: Stanford University Press, 2013.

Goree, Robert. "*Meisho Zue and the Mapping of Prosperity in Late Tokugawa Japan*," in *Cross-Currents. East Asian History and Culture Review* 23 (2017): 73–107.

Goree, Robert. *Printing Landmarks: Popular Geography and Meisho Zue in Late Tokugawa Japan*. Cambridge, MA: Harvard University Asia Center, 2020.

Haft, Alfred. *Aesthetic Strategies of The Floating World: Mitate, Yatsushi, and Fūryū in Early Modern Japanese Popular Culture*. Leiden: Brill, 2012.

Horiuchi, Annick, and Daniel Struve eds. *Guide illustré des produits renommés des monts et mers du Japon. New edition*. Paris: Collège de France, 2020.

Kikuchi Isao 菊池勇夫. "*Ryūminzu ni miru Tōhoku Daikikin* 流民図にみる東北大飢饉," *Shiroi Kuni no Uta* 白い国の詩10 (2004): 4–11.

Kinski, Michael. "Boxes, Fabrics, and Mirrors: On the Contents and the Classification of Popular Encyclopedias from Early Modern Japan," in Annick Horiuchi and Matthias Hayek eds., *Listen, Copy, Read: Popular Learning in Early Modern Japan*. Brill, 2014, pp. 70–88.

Latour, Bruno. *Politics of Nature: How to Bring the Sciences into Democracy.* Cambridge, MA: Harvard University Press, 2004.

Lippit, Yukio. *Colorful Realm: Japanese Bird-and-flower Paintings by Itō Jakuchu*. Chicago, IL: University of Chicago Press. 2012.

Marcon, Federico. "All Creatures Great and Small: Tokugawa Japan and its Animals," in Robert T. Singer and Kawai Masatomo, eds., *The Live of Animals in Japanese Art*. Princeton, NJ, and Oxford: Princeton University Press, 2019, pp. 29–37.

Marcon, Federico. *The Knowledge of Nature and the Nature of Knowledge in Early Modern Japan*. Chicago, IL, and London. Chicago University Press, 2015.

Oka, Midori. "*Nanga*: Chinese-Style Landscapes and Literary Poetics," in John Carpenter ed., *The Poetry of Nature: Edo Paintings from the Fishbein-Bender Collection*. New Haven, CT and London: Yale University Press, 2019, pp. 81–116.

Osaka Museum of History 大阪歴史博物館. *Kimura Kenkadō: Naniwa chi no kyojin. Tokubetsuten botsugo 200-nen* 木村蒹葭堂: なにわ知の巨人: 特別展没後200年記念. Kyoto: Shibunkaku shuppan, 2003.

Screech, Timon. *The Western Scientific Gaze and Popular Imagery in Later Edo Japan: The Lens within the Heart*. Cambridge: Cambridge University Press, 1996.

Shirane, Haruo. *Japan and the Culture of the Four Seasons: Nature, Literature, and the Arts*. New York: Columbia University Press, 2012.

Steininger, Brian. *Chinese Literary Forms in Heian Japan: Poetics and Practice*. Boston, MA: Harvard University Asia Center, 2017.

Thompson, Sarah E. *Hokusai's Lost Manga*. Boston, MA: Museum of Fine Arts Boston, 2016.

Yokota Fukuhiko 横田冬彦, *Dokusho to Dokusha* 読書と読者. Tokyo: Heibonsha, 2015.

Zwicker, Jonathan. "Playbills, Ephemera, and the Historical Imagination in Nineteenth-Century Japan," *The Journal of Japanese Studies*, 35.1 (2009): 37–59.

About the Author

Doreen Mueller is university lecturer of Japanese art and material culture at Leiden University. She obtained a PhD in the History of Art from SOAS, University of London. Her research explores visual culture's intersections with social and environmental history, focusing on how representations of famine and natural disasters were used for moral admonition and social criticism in early modern Japan. She has published on the production of knowledge about animals and plants in early modern East Asia, and the nature of political caricature in nineteenth-century Japan.

2. Tea Harvesting at Uji: Repackaging Uji as a Productive Place

Shiori Hiraki

Abstract: This chapter investigates the production and the dissemination of knowledge about a particular region in early modern Japan, namely, late-seventeenth- to early-eighteenth-century Uji. Uji is an interesting case study that shows how traditional knowledge was being negotiated with the new. Taking a scroll, *Tea Harvesting at Uji* by Kanō Tansetsu as its focus, this chapter argues that the traditional pictorial language of Uji was manipulated to promote the image of Uji as a productive tea region peacefully governed by the Tokugawa Shogunate.

Keywords: Tea, Uji, famous place, poetic pillow, Kanō Tansetsu

Introduction

The scroll opens a little abruptly, but it works: the people crossing the bridge immediately invite the viewer to 'enter' the scroll and follow them.[1] We see boats with boatmen and passengers on board. The men on the large boat are drinking a cup of sake, and the children are trying to catch fireflies. The fresh green willow leaves and the fireflies indicate that the season is late spring or summer. In front of the house, at the foot of the bridge, a man sits chatting with someone inside, possibly the master of this house. He is pointing towards a man and a woman, presumably husband and wife, with two children. His gesture, and that of the woman, carry the viewer further into the scene. The painter of this scroll often uses this kind of device to encourage the viewer to move on. We see a hut by the river where a water wheel is running. Bales of rice have been stored in one of its rooms, indicating that

1 I would like to thank Fan Lin and Doreen Mueller at Leiden University for their support, and Elesabeth Woolley for giving me an important piece of advice on translations. Of course, all errors remain my own. My sincere thanks go to Mary Redfern (then Curator of the East Asian Collections at the Chester Beatty Library, Dublin) for organising the viewing of the work.

Fan Lin and Doreen Mueller (eds), *Picturing Animals and Plants in Early Modern China and Japan. Innovation, Experiments, and Anxieties*. Taylor & Francis Group, 2025
DOI: 10.5117/9789048559091_CH02

the rice harvest has just finished. Following the river, we encounter a group of men and women engaged in washing and drying cloth, a typical image of bleaching cotton. Here, too, a movement towards the left is created by the faces and bodies of the people, as well as by the river and the lengths of drying cloth. At the end of this part, a child points towards a pine tree further to the left, which somehow indicates a change of scene. Two men run along the other side of the pine tree, one of them pointing towards what is probably their destination: a field of tea trees where men, women, and children are picking tea leaves. The lively and pleasant atmosphere is underlined by the painter's sensitive use of multiple green colours and by the delicate patterns and hues of the labourers' attire, which is too luxurious for picking leaves (and bleaching cotton). Continuing our walk, we find that the picked leaves are carried into a house where tea jars for storage are ready. A pine tree and another tree (perhaps a tea tree) conclude the scroll. The entire scroll is framed by two streams of clouds, at the top and bottom of the image, as if to give the viewer a sense of peeping into another world.[2]

The place the painter visualised in this scroll is Uji, famous for tea production since the fifteenth century. Based on the signature at the end of the scroll, the painter is believed to be Kanō Tansetsu (1655–1714), an artist of the famous Kanō school, which served the Tokugawa shogunate during the Edo Period (1603–1868). Tansetsu was the second son of Kanō Tan'yū (1602–1674), who had established the school in Edo and entered the service of Tokugawa Ieyasu (1542–1616), the founder of the Tokugawa Shogunate. The first researcher to identify the place in Tansetsu's scroll as Uji was Kōno Motoaki, an expert on early modern Japanese painting. In his discussion of this scroll in volume five of *Japanese Art: The Great European Collections* (Hizō Nihon bijutsu taikan), Kōno explained the scenes with the help of *Illustrations of Famous Places of the Capital* (Miyako meisho zue).[3] The scroll does not have any text, but Kōno convincingly identifies the bridge with a terrace projecting over the river as Uji Bridge, which connected the eastern and western banks of Uji River in the Edo period. Together with the depictions of tea harvesting and cotton bleaching, two of the most famous industries in the area, the identification of the place as Uji is logical.

Kōno wrote only a short commentary and used *Illustrations of Famous Places of the Capital* merely to identify the location depicted in the scroll. Robert Goree's recent analysis of *Illustrations of Famous Places of the Capital* demonstrates that it was not published to convey reality but a sense of prosperity and peace, to be used

2 The image is available on the collection website of the Chester Beatty Library: https://viewer.cbl.ie/viewer/image/J_1137/3/LOG_0000/.

3 Hirayama Ikuo and Kobayashi Tadashi eds., *Chestā Bītī Raiburarī*, vol. 5, Hizō Nihon bijutsu taikan (Tokyo: Kōdansha, 1993), p. 264.

as virtual tours by 'armchair travellers.[4] Moreover, *Illustrations of Famous Places of the Capital* was published in 1780, almost a hundred years after Tansetsu executed his scroll. It can be argued that what Tansetsu visualised became the precursor of the image of Uji as seen in later published books, including *Illustrations of Famous Places of the Capital*.

Since the body of Tansetsu's works and the seals he used remain to be researched, it is difficult to determine when the scroll was painted. Some characteristics seem to indicate that he painted it in his youth: his frequent use of two people, the right one facing right and the other left, to generate movement of the eye towards the left; the lack of coherence in the pose of bodies, and the ink lines executed with a fragile touch. Tansetsu might still have been in training – there is not enough information to situate this work within his oeuvre. In addition, neither the scroll nor the box containing it provide any information as to who commissioned the scroll and why. Therefore, this paper will focus on an art historical analysis of the scroll to find out how Tansetsu created the image. This orthodox method is helpful in the case of this scroll, because it belongs to the earliest group of works representing the Uji tea harvest and thus the adaptation of the traditional pictorial language to create a new image of Uji is an important question. Uji was a *famous place* (*meisho*) renowned for its religious connotations and poetic resonances created by *waka* poetry since ancient times, and the tea industry was rather a new theme in the area. Investigating Tansetsu's work means exploring the creation of a new image of a *famous place* that underwent a radical change in the early modern period. I will use some books published during the period when the artist was active, not so much to position the scroll within its historical period as to clarify its unique position in the diachronic formation of the image of Uji.[5]

There are several works depicting tea production in Uji whose painters have been identified. One is the screen on the right of *Ritual of Racehorse at Kamigamo Shrine and Picking Tea at Uji* by Kusumi Morikage (dates unknown)[6] (Figure 2.1) and the

4 Robert Goree, *Printing Landmarks: Popular Geography and Meisho Zue in Late Tokugawa Japan* (Cambridge, MA: Harvard University Asia Center, 2020), pp. 33–68.

5 However, this does not mean that there are no possible clues. In 1691, when the fifth Tokugawa shogun, Tokugawa Tsunayoshi (1646–1709), visited his trusted advisor Yanagisawa Yoshiyasu (1658–1714) at his residence in Edo, Yoshiyasu presented the shogun with a scroll of tea harvesting at Uji executed by Tansetsu. The event is mentioned in Yoshiyasu's biography, *Annals of the Hall of Rejoicing in Virtue* (Rakushidō nenroku), but apart from the title and the name of the painter no details are given. See Miyakawa Yōko ed., *Rakushidō nenroku*, vol. 1 (Tokyo: Yagi Shoten, 2011), p. 62.

6 These screens are usually paired, but there has been no research as to whether these screens were intended as a set. Kusumi Morikage trained under Kanō Tan'yū, but he left the school possibly due to the bad behaviour of his son. Morikage worked for the Kaga domain in the Hokuriku region after leaving Edo.

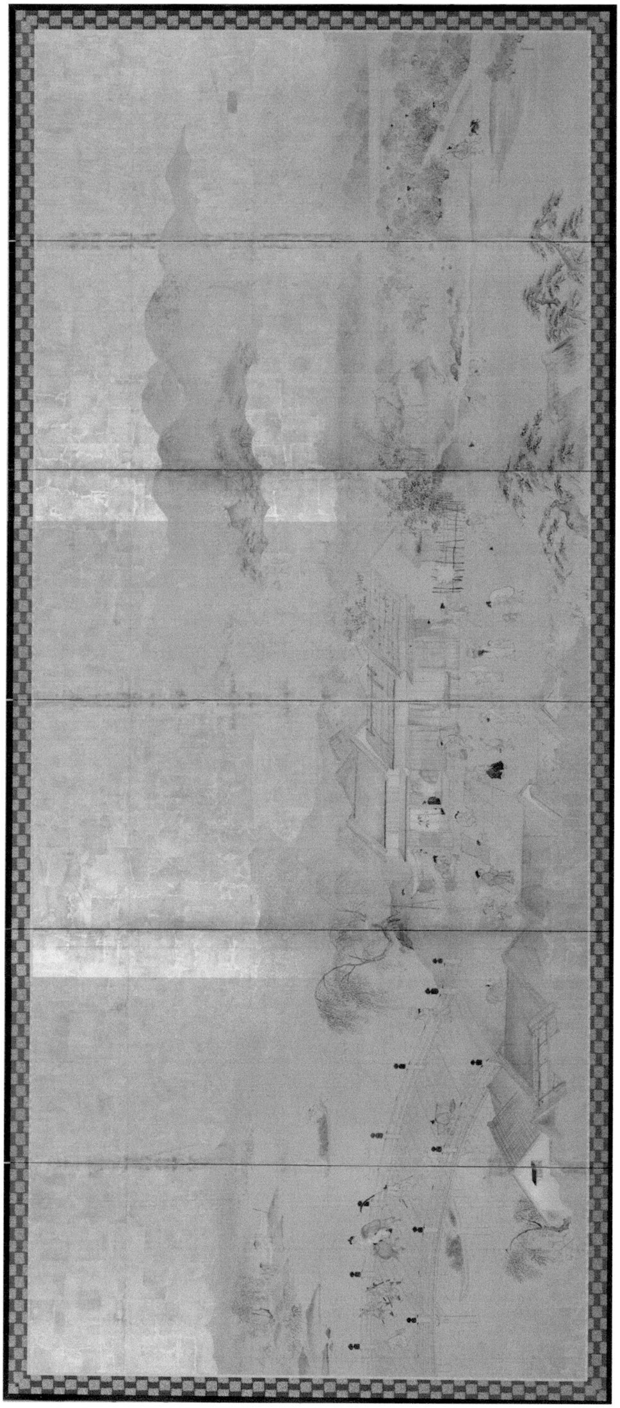

Figure 2.1: Kusumi Morikage, *Picking Tea at Uji*. One of a pair of six-panel folding screens, *Ritual of Racehorse at Kamigamo Shrine and Picking Tea at Uji*. 17th–18th century. Ink and colour on paper, 151.5 x 362.5 cm. Right screen pictured. Important Cultural Property. Okura Museum of Art, Tokyo.

other is a scroll of *Scenes of Tea Production in Uji* by Kaihō Yūsen (1690–1741);[7] both were executed in the late seventeenth or early eighteenth century. Tansetsu's work is very different from these because he did not include the religious institutions that were often depicted in representations of Uji and instead inserted some scenes that were rarely visualised in creating the image of the area. In addition, his focus on the highlights of the depicted industries suggests that Tansetsu's scroll was not meant to provide instruction on the production processes. My art historical analysis will reveal that Tansetsu intended his scroll to show a harmonious, human-centred Uji where everyone had a share in the industry. To create such an image, Tansetsu removed religious references and changed poetic symbols to pragmatic ones. This shift in visuality, where an antique form of representation fostered through the medieval period was breaking down and modernity was emerging, is the focus of this scroll.

From Religious to Secular

Before starting the analysis of this scroll, let us briefly look at the geographical circumstances of the area in question. Uji is located about eight kilometres south of Kyoto. The area is divided by Uji River running from southeast to northwest. Uji Bridge connects its eastern and western banks (Map 2.1). On the western bank is Byōdō-in (Temple of Equanimity), converted from a villa into a Pure Land Buddhist temple by Fujiwara no Yorimichi (992–1074) in 1052.

On the eastern bank, to the north, we find Manpuku-ji (Temple of Ten-Thousand-Fold Happiness), a temple of the Ōbaku Zen sect founded by Yinyuan Longqi (J: Ingen Ryūki, 1592–1673) in 1661. From Manpuku-ji, we can see Uji River curve towards the west and merge with two other rivers, the Kizu and the Katsura. It continues to Osaka Bay as the Yodo River.

Over the centuries, these highly important places of worship became landmarks within the region – making Uji one of those *famous places* one had to visit. These *famous places* were usually visualised with iconic motifs in the region and other motifs and references portrayed in classical literature, as will be discussed in detail later. In the case of Uji, Byōdō-in was one such motif to represent the area. The temple seen from the eastern bank over the bridge facing west epitomised the visualisation of the Pure Land. Indeed, the earliest image of Byōdō-in on record

7 Yūsen was the great-grandson of Kaihō Yūshō (1533–1615), a painter who worked under Kanō Eitoku (1543–1590), the Kanō master who expanded the business of the school through commissions from notable warlords such as Oda Nobunaga (1534–1582) and Toyotomi Hideyoshi (1536–1598). For the illustration, see Uji-shi Rekishi Shiryōkan, ed., *Uji cha: meisho e kara seicha zu e* (Uji: Uji-shi Rekishi Shiryōkan, 1985), pp. 40–41.

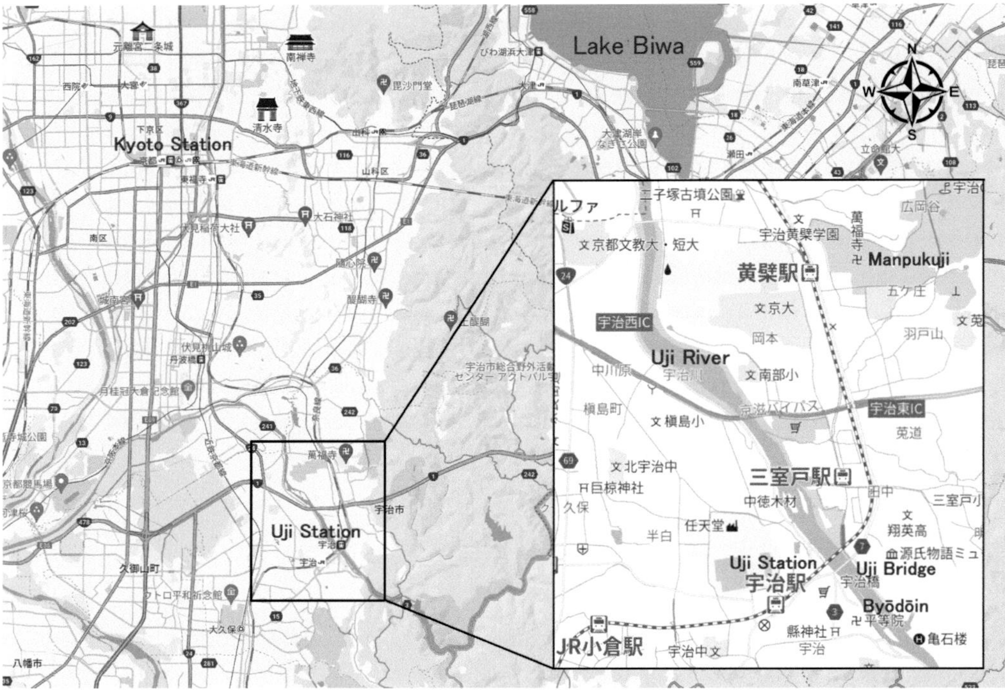

Map 2.1: Map of Uji.

was the wall painting on the west wall of the Amida Hall of the Kanjizaiō-in in Hiraizumi, which was recorded in Documents of the Chūson-ji Sutra Repository from 1313.[8] As art historian Matthew McKelway has pointed out, the painting on the western wall of the Amida Hall had a double meaning: Kyoto was located to the west of Hiraizumi, the same direction as the Pure Land Western Paradise of Amida Buddha.[9] Other visual works representing Uji as a religious site usually depict Byōdō-in as seen from the eastern bank, too. One example is a pair of four-panel folding screens titled *Famous Places Outside the City* by Kanō Eitoku (1543–1590), which was probably made around 1565.[10]

With this geographical knowledge in mind, let us go back to the opening scene of the scroll (Figure 2.2). What immediately draws the viewer's attention is the terrace in the middle of the bridge that juts out over the river. This characteristic terrace is called the *san-no-ma*, which literally means 'the third space,' indicating the space between the third and fourth pillars from the west end (hereafter referred to as the

8 Tsuji Nobuo, *Rakuchū rakugai zu*, vol. 121, *Nihon no bijutsu* (Tokyo: Shibundō, 1976), p. 21.

9 Matthew P. McKelway, *Capitalscapes: Folding Screens and Political Imagination in Late Medieval Kyoto* (Honolulu: University of Hawai'i Press, 2006), p. 30.

10 Kano Hiroyuki, "Kanō Eitoku hitsu Rakugai meisho yūraku zu byōbu (Uji-gawa, Ōi-gawa)," *Kokka*, no. 1331 (2006): 22. For the illustration, see Plate 2 of this article.

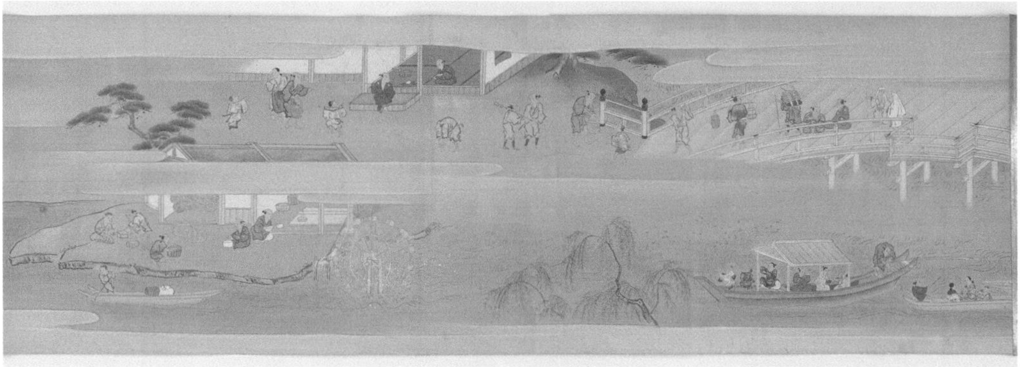

Figure 2.2: Kanō Tansetsu Morisada, *Tea Harvesting at Uji*. 18th century. Handscroll, ink and colour on paper, 32 x 519 cm. Chester Beatty Library, CBL J 1137. Part of the scroll pictured. © The Trustees of the Chester Beatty Library, Dublin.

san-no-ma terrace). The terrace was constructed in the early modern period, based on the story that warlord Matsunaga Hisahide (1510–1577) drew water from here to host the famous tea master, Sen no Rikyū (1522–1591), during his tea gathering in 1565, and the legend that it had been the location of a shrine dedicated to the Goddess of the Bridge.[11] Some of the earliest representations of this terrace may be found in *The Illustrated Tale of Genji* (Eiri Genji monogatari) published in 1650 with illustrations by Yamamoto Shunshō (1610–1682).[12] The *san-no-ma* terrace was built facing southeast, upstream, and it is depicted like this in other works, such as the *Scene of Uji* by an anonymous painter, which was probably executed between 1668 and 1678[13] (Figure 2.3) and Kaibara Ekiken's *Excellent Views of Kyoto* (Keijō shōran)[14], published in 1706. Since the viewer is facing this terrace in Tansetsu's scroll, the left bank must be the western bank. Remarkably, however, the buildings that identify it as the western bank, Byōdō-in and the Hashihime Shrine, are missing. The Hashihime (Lady of the Bridge) Shrine, where one could worship the Goddess of the Bridge, was located at the bridge's western foot, as can be seen in the *Screen of Uji* and in *Excellent Views of Kyoto* (see 'Hashihime no miya' to the left of Byōdō-in). The two sites also already appear on the screens *Famous Places*

11 Uji-shi Rekishi Shiryōkan, *Uji-bashi: sono rekishi to bi to* (Uji: Uji-shi Rekishi Shiryōkan, 1995), pp. 54–55.

12 Uji-shi Rekishi Shiryōkan, *Uji-bashi: sono rekishi to bi to*, pp. 54–55.

13 Noguchi states that if it can be supposed that representations reflect reality, it is possible to situate this work after 1668, when the main halls of Manpuku-ji were constructed, and before 1678, when the *sanmon* (the gate with three spaces between the pillars) was built. Noguchi Takeshi, "Uji zu byōbu," *Kokka*, no. 1392 (2011): 28.

14 For the illustration, see the digitised images in the Waseda University Library, https://archive.wul.waseda.ac.jp/kosho/ru04/ru04_05298/ru04_05298_p0045.jpg and https://archive.wul.waseda.ac.jp/kosho/ru04/ru04_05298/ru04_05298_p0046.jpg.

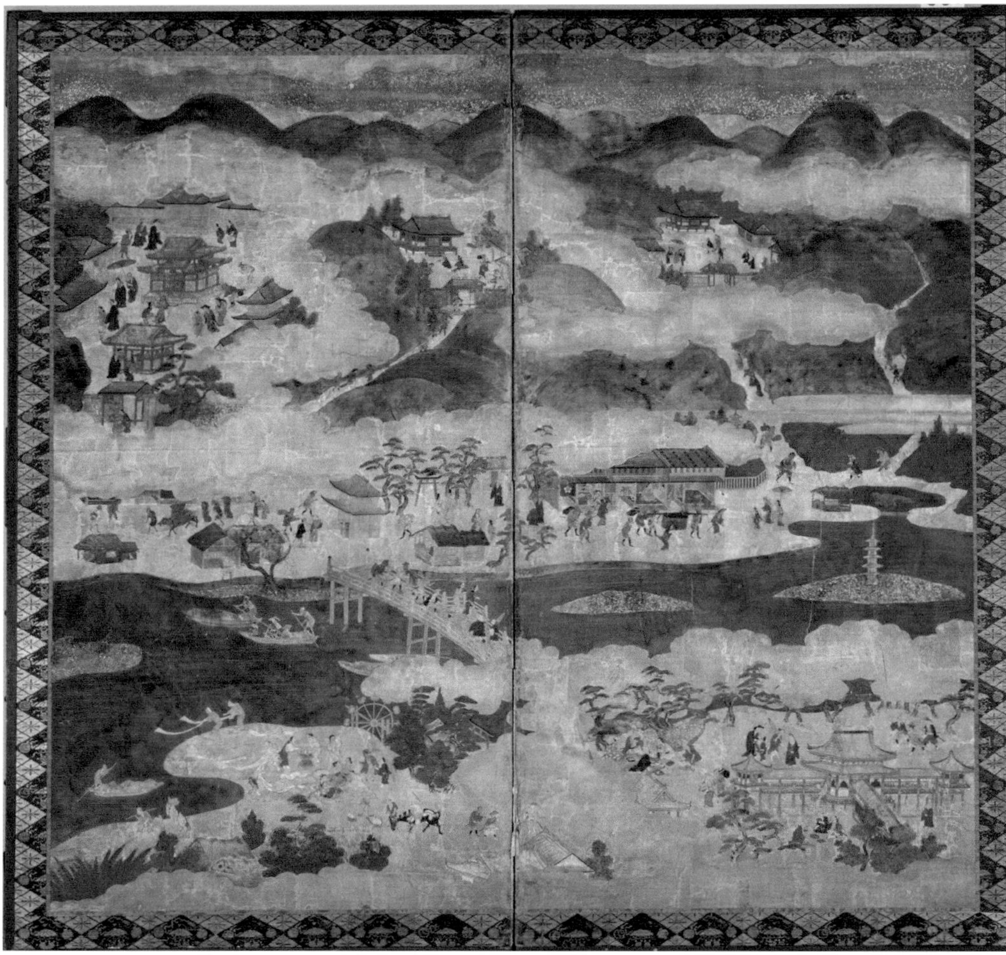

Figure 2.3: Anonymous, *Scene of Uji*. 17th century. Two-panel folding screen, ink and colour on gold-foiled paper, 146 x 157.8 cm. The Nezu Museum, Tokyo.

Outside the City by Kanō Eitoku.[15] It is not known whether Tansetsu made use of these earlier works and published books, but it is reasonable to assume that he was aware of the presence and the location of these two sites.[16]

15 See Plate 2 of Kano, "Kanō Eitoku hitsu Rakugai meisho yūraku zu byōbu."

16 The painters, including those of the Kanō school created the images on the basis of copies transmitted through generations. For the school's emphasis on training over talents, see Karen M. Gerhart, "Talent, Training, and Power: The Kano Painting Workshop in the Seventeenth Century," in Brenda G. Jordan and Victoria Weston, eds., *Copying the Master and Stealing His Secrets: Talent and Training in Japanese Painting* (Honolulu: University of Hawai'i Press, 2003), pp. 18–29. For the details of training based on copybooks, see Brenda G. Jordan, "Copying from Beginning to End? Student Life in the Kano School," in *Copying the Master and Stealing His Secrets: Talent and Training in Japanese Painting,* pp. 33–58.

It should also be mentioned that the image of the right-hand side of the Uji Bridge may have been cut off at some point. The depiction of only the prow of a boat emerging seems to be an incomplete composition for the beginning of a scroll. In that case, it would be interesting to know what was depicted on the eastern bank. However, let us leave this aside and focus on what Tansetsu is doing here. Instead of the Hashihime Shrine, he depicts a house where a man is being served a drink[17] and points towards a couple and (presumably) their children, giving a glimpse of local people's everyday life. On realising this, the viewer becomes aware of the idea of Uji that Tansetsu is trying to convey. He is downplaying the religious meanings of the place while highlighting its vernacular aspects as a place for conducting daily business.

Two men are sitting on the bridge, others are crossing it from right to left. A man and a woman wearing white robes and a man wearing a hat are travellers who have probably come to visit Uji's iconic religious sites; other indications are their travellers' staffs and the black leggings (*kyahan*) worn by the man with the hat. These figures resemble those appearing on earlier works showing the passage on the bridge. Since there are no religious sites in this scroll, we can only guess their destination, but what is remarkable is the attention they get. The man sitting on the right and the man carrying the baskets are both looking at the man with the hat, and the man looking back while lifting his left hand is taking in all three travellers. In this way, we are reminded that Uji is a famous destination, even though the actual sites are nowhere to be seen. The important point is that the attention the travellers get sets them apart from the locals and their daily chores.

The role of the bridge in this image is to show that this place is none other than Uji and to create a one-way passage towards the western bank, where people engage in their daily business. The representation of the bridge here lacks the lively comings and goings between the religious sites of the western and eastern banks seen in the works by previous painters. The two men at the foot of the bridge looking at each other may also be travellers; they are probably the attendants of a person of higher status, because one of them is carrying a *hasami bako* on his shoulder, a travelling

17 Kōno speculated that this house is the tea house run by a master called Tsūen (dates unknown), as explained in *Illustrations of Famous Places of the Capital*, but whether it is represented in this scroll remains in doubt. It is introduced in *A Kyoto Souvenir from Dekisai*, published in 1677, as a tea house at the foot of Uji Bridge on the eastern bank. See Asai Ryōi, "Dekisai Kyō miyage," in Noma Kōshin ed., *Shinshū Kyōto sōsho*, vol. 11 (Kyoto: Rinsen Shoten, 1976), p. 651. In 1793, the Tsūen family submitted a record of family history titled *A Record of Miscellaneous Matters Regarding the Uji Bridge* (Uji bashi shoji todome ki) to the shogunate, claiming that their job included the prevention of fire and the formation of pools of water in case of heavy rain, and the supervision of the passage of the procession of tea jars and visiting shogunal inspectors. For this reason, the family used lanterns with the inscription *goyō* (literally 'official business'). The record was kept updated until 1850. See Uji-shi Rekishi Shiryōkan, *Uji-bashi: sono rekishi to bi to*, pp. 35–37.

chest used to carry the belongings of high-ranking members of the warrior class. It is not clear whether they are about to cross the bridge or are simply waiting for their master. No one stands on the *san-no-ma* terrace to watch the upstream view. Again, Uji does not attract visitors as a site of religious institutions in this scroll.[18]

In earlier representations of Uji (for example, in Eitoku's screen), boats were often transporting brushwood. Tansetsu, however, stresses the aspect of merrymaking. The people in the boats are clearly enjoying themselves. Whether they are also travellers who have come for sightseeing is unclear, but the fact that Tansetsu presents Uji River as a place where people can enjoy themselves pushes the religious aspect of the region even further into the background. His representation of children catching fireflies may be the first to depict this motif in the history of representing Uji. According to a chronological list of the representations of Uji made by art historian Takeuchi Misako, fireflies appear only once, in *The Infinite Treasury of Plum Blossoms* (Baika mujinzō), a collection of poems by Rinzai Zen monk Banri Shūku (b. 1428) in the late fifteenth century: his poem was the inscription of a painting of fireflies at Uji.[19] In the poem, Banri Shūku compares the light emitted by fireflies to the flames of war in ancient China, based on Uji's history as the battlefield of the Genji and Taira clans.[20] The comparison between the fireflies and the souls of the deceased is still seen in some gazetteers, for example in Asai Ryōi's (d. 1691) *A Kyoto Souvenir from Dekisai* (Dekisai Kyō miyage).[21] Given this long history of comparison, Tansetsu's affectionate eye on children who are absorbed in catching fireflies is remarkable. It is possible that, by employing children, he foregrounded their enjoyment of the fireflies as natural things without alluding to their literary and historical meanings. The painter deliberately changed the current of the river to show this scene effectively. As the *san-no-ma* terrace is facing upstream, the current should run from the front to the back of the image. Instead, Tansetsu sectioned the scene with a mist of gold powder and positioned the boats beneath the mist, so that they would follow the movement of the scroll, from right to left, thereby going against the current of the river.

To sum up, unlike earlier works, Tansetsu's scroll does not show Uji as a religious site. There are no religious institutions for travellers to visit. The Uji Bridge carries people to the left bank where others are conducting their daily affairs. The house

18 Noguchi suggests that some of the depictions in *Scene of Uji* suggest a decline in the sacredness of the west bank, such as the men feasting near Byōdō-in and a mother holding her child and talking with other women in Makinoshima. Noguchi, "Uji zu byōbu," 27.

19 Takeuchi Misako, "Ryūkyō suisha zu byōbu shinshutsu bon no shōkai o kanete jō," *Kokka*, no. 1138 (1990): 28.

20 Ichiki Takeo, trans., *Baika mujinzō chūshaku*, vol. 1 (Tokyo: Zoku Gunsho Ruijū Kanseikai, 1993), pp. 20–21.

21 Asai Ryōi, "Dekisai Kyō miyage," pp. 651–652.

at the foot of the bridge looks like a tea house rather than the Hashihime Shrine. Travellers attract attention as markedly different beings from the locals. People enjoying their boat ride and children catching fireflies add a merry atmosphere to the scroll. All this indicates that Tansetsu's scroll is based on a different scheme than Eitoku's screen and the anonymous *Scene of Uji*.

From Poetic to Practical

Along the river we see a waterwheel in operation, as well as a hut with two rooms; the left room serves as storage for rice bales and the right room has a kind of apparatus inside, which may relate to the waterwheel, although the mechanism is rather puzzling. The current here is the opposite of that in the opening scene; the river flows from left to right as is clear from the waterwheel rotating counterclockwise, and a boat moves towards the right.

The details have been simplified to such an extent that it is hard to identify all the tools and utensils, but the square box held by a man on the left-hand side on the bank is a *masu*, which is an implement for measuring rice. The rice is scooped into a tub of the type held by the man in the middle, who wears a brown kimono. With the help of these measuring tools, they fill rice bales holding about seventy-two litres of rice.[22] They seem to have finished their work and are putting away their tools. A man on the right holding a flask is already pouring a cup of sake to celebrate the completion of the rice storage. Behind him is a thresher, a tool that is usually set in motion by pedalling. However, Tansetsu seems to suggest that the waterwheel is the thresher's source of power, although the depicted structure is doubtful. The waterwheel is too far from the riverbank, and the role of the pipe that connects the box in the river with the hut is obscure. If the waterwheel was moving the thresher, it should be closer to the hut, and its axle should be attached to the thresher in some way to maximise the power of the water.

A waterwheel is one of the iconic features of Uji River. Uji was so famous for producing waterwheels for irrigation that the monk Kenkō (1283–1350) mentioned them in his famous *Essays in Idleness* (Tsurezuregusa).[23] Although there is a practical side to the relationship between Uji and waterwheels, most of the association comes from poetry. In the poetic tradition, the waterwheel was an indispensable part of Uji as a symbol of grief – the sadness of life being as constant as the turning

22 One *masu* holds about 1.8 litres (1 *shō*) of rice, and ten *shō* of rice (around 18.039 litres) will fill a tub. Four tubs would be enough for one rice bale.

23 Yoshida Kenkō, "Tsurezuregusa," in *Shinpen Nihon koten bungaku zenshū*, trans. Nagazumi Yasuaki, vol. 44, (Tokyo: Shōgakukan, 1995), p. 121.

of the wheel; primarily, a poetic play on the words Uji and *ushi*, meaning sad, gloomy.[24] A poem by the Buddhist monk Kisen (dates unknown) in *Collection of Waka Poems, Ancient and Modern* (Kokin wakashū), from the early tenth century, reads as follows:

> My dwelling is a hut, southeast of the capital, and I live quietly there:
> People say, I hear, that it is because I found the world distasteful that I came to live on Ujiyama, the mountain of grief.[25]

Art historian Tamamushi Satoko speculates that the belief in Pure Land Buddhism among the aristocracy in the eleventh century, characterised by the completion of Byōdō-in in 1053, further enhanced the image of Uji as a place of sadness and uncertainty and gives an example of a *waka* poem in *The Collection of Poems Compiled by Fujiwara no Sadayori* (Sadayori shū) in the early eleventh century, which employs the motifs of a waterwheel and Uji/*ushi* to represent this sentiment:

> I lament over the world, as I see a waterwheel by Uji River running,
> My sleeves are wet with tears.[26]

As these poems show, due to its association with sadness, Uji became a 'poetic pillow' (*utamakura*). Poetic pillows are a special category of place names and allusive terms used in *waka* poetry to create a particular poetic sentiment. They might introduce a specific physical feature of the place mentioned, real or imagined, or allude to connections with other forms of literature or historical events, or, alternatively, be used for wordplay to layer multiple meanings.[27] Poets used poetic pillows to create a web of references connecting their own poem with poems of the past, and a web of allusions associated with the pillow. In the case of Uji, Uji River was a source of motifs for producing webs of allusions, such as the bridge, with its associations with coming and going and thus with parting.[28] Willow trees were another motif; their association with bridges created a sense of parting in poems in Tang Dynasty China (618–907), which were probably

24 Edward Kamens, *Utamakura, Allusion, and Intertextuality in Traditional Japanese Poetry* (New Haven, CT: Yale University Press, 1997), p. 131, see also note 31 on pp. 275–276.

25 Kamens, *Utamakura, Allusion, and Intertextuality in Traditional Japanese Poetry*, p. 130.

26 Tamamushi Satoko, "'Ryūkyō suisha zu' to 'Uji no kawase no mizuguruma': kayō to fūryū to byōbu e no kōkyō," in *Nihon bungei shichōron* (Tokyo: Ōfūsha, 1991), pp. 865–866.

27 Kamens, *Utamakura, Allusion, and Intertextuality in Traditional Japanese Poetry*, p. 28.

28 Timon Screech, *Obtaining Images: Art, Production and Display in Edo Japan* (London: Reaktion Books, 2012), p. 63.

introduced to Japan in later years.[29] An example that became widely known today as *Screens of Bridge and Willow*, crystallises these layered literary references.[30]

In Tansetsu's scroll, however, no sense of sadness can be detected. The people on Uji Bridge and Uji River are working or enjoying themselves. The waterwheel is simply a tool, without any connotation of gloominess. This, again, marked a radical change in the use of motifs in images. The waterwheels in Uji failed to capture people's attention, possibly because warlord Toyotomi Hideyoshi (1536–1598) had cleared Uji Bridge in 1594 during the construction of Fushimi Castle. The bridge was reportedly reconstructed by Tokugawa Ieyasu in 1599, but it seems that five years were enough for people keen on fashionable lyrics and songs to forget about it. Instead, the waterwheel in Yodo River started to feature in a song performed by *kabuki* players. The song alluded to love by comparing a waterwheel to one half of a couple waiting for the other.[31] Although the waterwheel in Uji disappeared from literature, it persisted in images, mainly because it indicated that the place was indeed Uji. Tansetsu's scroll not only used this iconic motif to represent the region, but it also removed literary references, turning the waterwheel into a practical tool for agriculture.

The completion of harvesting and storing rice bales and the man pouring a cup of sake from a flask in celebration of this add a merry atmosphere to the scene. Rice was an important crop, not only because it was a staple food, but also because rice production formed the basis for taxing the domains. Thus, rice cultivation scenes became part of the imagery admonishing military elites to practice moral conduct; regional productivity depended on their protection and good governance. The theme of rice cultivation often comprised the procedure from beginning till end, from planting rice seedlings to storing grain. The scenes would also be set in an imaginary place, representing an ideal world governed by just rulers.[32] As Kōno

29 Adachi Keiko, "Nishu no ryūkyō zu byōbu," *Kokka*, no. 1135 (1990): 7, 10–11. Screech states that willow trees were included not only because these were common on riversides, but also because they reminded people of women's dishevelled hair, and thus of women who were sad and lonely, as read in the poem by Fujiwara no Teika (1162–1241). Screech, *Obtaining Images: Art, Production and Display in Edo Japan*, p. 65.

30 For the illustration, see Takeda Tsuneo et al., *Keibutsu ga: shiki keibutsu*, ed. Daiichi Shuppan Sentā, vol. 9, Nihon byōbu e shūsei (Kōdansha, 1977), pp. 76–79.

31 Tamamushi, "'Ryūkyō suisha zu' to 'Uji no kawase no mizuguruma': kayō to fūryū to byōbu e no kōkyō," p. 874.

32 Screech, *Obtaining Images: Art, Production and Display in Edo Japan*, pp. 149–151. It is said that the images of rice cultivation and silk production scenes as admonition were originally created in China by Lou Shou (1090–1162), a magistrate of Wuqian in Linan, who created twenty-one images of rice production and twenty-four images of silk manufacture, each with a poem explaining the scene, to describe the life of farmers under good governance, especially the proper conduct of local officials. Lou Shou presented the set to Emperor Gaozong (r. 1127–1162) of the Southern Song dynasty. Roslyn Lee Hammers, *Pictures of Tilling and Weaving: Art, Labor and Technology in Song and Yuan China* (Hong Kong: University Press, 2011), pp. 41–60. Although the set presented to the emperor no longer exists, a copy published as a printed

mentions, the relation of this scene to Uji is not obvious.[33] Indeed, the storage of rice is difficult to relate to an admonitory context as it is the only scene relating to rice cultivation in Tansetsu's scroll. A more plausible suggestion would be that it was meant to show Uji as 'countryside.' Adding people and things unrelated to the theme of an image is a well-known method to amplify its sense of reality. This can already be seen in the screen of Uji by Eitoku. To the right of the Hashihime Shrine, we see a man and his ox at work in the rice field, providing a glimpse of daily life in a screen in which Byōdō-in plays the central role.[34] In Tansetsu's scroll, too, men at work indicate a rural area, with the waterwheel identifying the region as Uji.

Most likely, this scene was based on copies rather than actual observation. Tansetsu may have used some images of rice production scenes that were current in the Kanō school. The appliance that looks like a thresher can be seen in a pair of six-panel folding screens by Kusumi Morikage (Figure 2.4), in which two men set their right feet on it. This appliance first appears in the woodblock-printed version of the *Pictures of Tilling and Weaving* published in Ming dynasty China in 1462 by Song Zonglu, reprinted in Japan in 1676 and illustrated by Kanō Einō (1631–1697), the eldest son of Kanō Sansetsu (1589/90–1651), the master of the Kyoto branch of the Kanō school. Little is known of the exchange of imagery between the Kyoto and Edo branches of the Kanō school; Tansetsu's scroll might indicate that Kanō painters in Edo used copies and printed books published by Kanō painters in Kyoto.[35] The inaccuracy in the structure of the waterwheel and the way in which the rice bales have been stored (they seem to have been thrown in carelessly, even though each bale would weigh around sixty kilos) can probably be explained by this circumstance.

One final problem to be solved is a break in the seasonal cycle: harvesting and storage are conducted in autumn, whereas the scenes before and after take place in summer. This suggests that a powerful scheme of poetic pillow was at work even in Tansetsu's time. Still, what the artist did was innovative. As mentioned, Uji was always set in autumn to reference a feeling of sadness. Tansetsu changed the

book by Lou's descendants became well known and eventually reached Japan. Reizei Tamehito et al., *Mizuho no kuni Nihon: shiki kōsaku zu no sekai* (Kyoto: Tankōsha, 1996), p. 4.

33 Kōno explains that these rice bales are for paying the annual tax. Kōno, the entry of this scroll in Hirayama and Kobayashi, *Chestā Bītī Raiburarī*, p. 264.

34 See Plate 2 of Kano, "Kanō Eitoku hitsu Rakugai meisho yūraku zu byōbu (Uji-gawa, Ōi-gawa)."

35 A pair of screens by Sansetsu in the collection of Tokyo University of the Arts also has this type of thresher. Reizei et al., *Mizuho no kuni Nihon: shiki kōsaku zu no sekai*, pp. 30–31. For details of Song Zonglu's *Pictures of Tilling and Weaving*, see Reizei et al., *Mizuho no kuni Nihon: shiki kōsaku zu no sekai*, pp. 24–25. Kōno Michiaki, a historian of agricultural tools, states that while the use of Song Zonglu's edition by the Kanō school painters in Kyoto is evident, this is not the case for those in Edo. Reizei et al., *Mizuho no kuni Nihon: shiki kōsaku zu no sekai*, p. 25. The interaction between the Kanō schools in Kyoto and Edo will be clearer if other images including this type of thresher in the works by the Edo Kanō painters are confirmed as based on Song's edition, and information on the commissioning process emerges.

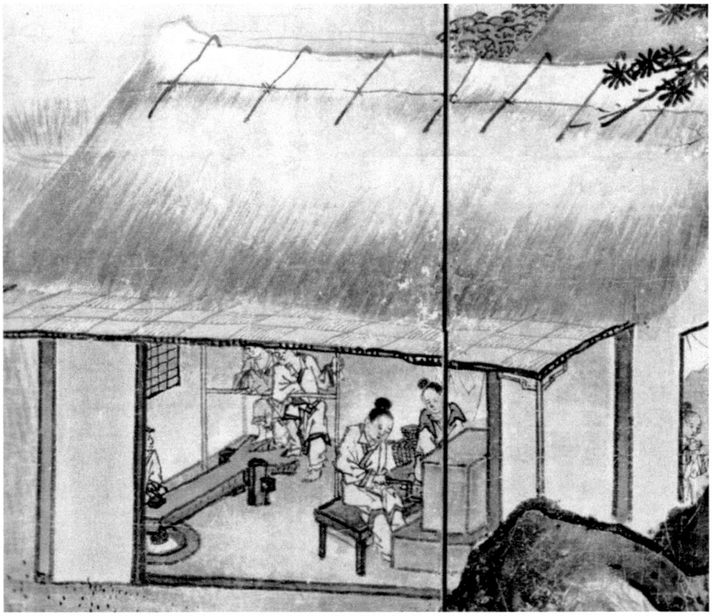

Figure 2.4: Kusumi Morikage, *Farming Scenes*. 17th century. Pair of six-panel folding screens, ink and light colour on paper, 155.7 x 345 cm each. Part of the first and second panels of the right screen pictured. Tokyo National Museum. Image: TNM Image Archives.

atmosphere of Uji to a place filled with cheerfulness by highlighting the region's productivity. A pine tree, a symbol of eternity due to their being evergreen, stands behind the rice storehouse and adds a celebratory tone to the scene. It even extends its branches leftward to guide the viewer to the next scene, working as a partition between two sections and as guidance from one part to the next. Taken as a whole, the scene shows Tansetsu's ingenuity in challenging the traditional scheme of depicting Uji by fully deploying traditional motifs.

Cotton Bleaching as Harmonious Labour

What comes next in Tansetsu's scroll is a cotton bleaching scene in Makinoshima village, located on the western riverbank in the northern part of the Uji region. Cotton bleaching was a famous industry in Uji; it featured in some early *waka* poems, but there are no details on the industry from the early modern period. It has been argued that embankment works ordered by Toyotomi Hideyoshi in 1594 caused the water level of Uji River to fall, leading to the decline of the cotton bleaching industry.[36] As *The Illustrated Encyclopedia of Humanity* (Jinrin kinmō zui, published in 1690) says:

36 Ashikaga Kenryō et al., "Uji-gawa no chisui," in *Chūsei no rekishi to keikan*, vol. 2, Uji-shi shi (Kyoto: Uji Shiyakusho, 1974), pp. 608–609.

> Cotton bleaching started in Makinoshima in Uji. In Kyoto, it was on the Gojō-gawara River. Nowadays, Nara produces the best quality of cotton.[37]

This implies that, by the late seventeenth century, the industry was no longer in full operation in the Uji region. The unrealistic way of drying cotton cloth in Tansetsu's scroll may, again, have been the result of the use of models in copybooks, from which the painter made an assemblage of the industry as imagined since ancient times. Eitoku's screen and a fan painting by his father, Kanō Shōei (1519–1592), likewise show the cotton bleaching industry by Uji River, and these depictions may have laid the foundations for envisioning the industry in that area (Figure 2.5). Although the industry itself no longer flourished, its importance for representing Uji did not diminish in the seventeenth and eighteenth centuries. This is evident from works by Morikage (Figure 2.1) and Yūsen,[38] which testify to the continued importance of the industry in signifying Uji even though this no longer accorded with contemporary historical realities. But did Tansetsu's scroll carry the same connotations?

Legend has it that cotton bleaching was taken up in Uji in 1284 on the instructions of a monk named Eison (1201–1290) from Saidai-ji Temple in Nara. When Eison was charged with supervising the construction of Uji Bridge, he, in accordance with Buddhist teaching, forbade the fishermen of Makinoshima village to fish and taught them cotton bleaching to make their living.[39] Depicting cotton bleaching thus became part of the imagery of sacredness that characterised Uji, especially when it was represented together with Buddhist institutions such as Byōdō-in, and the thirteen-storeyed rock pagoda tower on a small island on the upper stream of Uji River, beneath which fishing tools and boats that had reached the end of their working lives were buried.[40] As has already been pointed out, however, Tansetsu's scroll does not include the religious institutions of the area.

The lack of relationship between the depicted scene and other historical and geographical characteristics of the area is strengthened by the mist that covers the top and bottom of the image, hiding most of the background. As a result, the viewer's eye is fixed on the figures who are harmoniously working together in finely decorated attire that is too beautiful for their work. Stripped of religious connotations, the cotton bleaching industry, though outdated, signified Uji, pinning down an image of productivity through labour in a peaceful atmosphere.

37 Asakura Haruhiko, ed., *Jinrin kinmō zui* (Tōyō bunko 519. Heibonsha, 1990), p. 197.

38 For Yūsen's work, see Uji-shi Rekishi Shiryōkan, *Uji cha: meisho e kara seicha zu e*, p. 8.

39 Fujii Manabu and Moriya Shigeru, "Shūkyō to hōjō shisō," in *Chūsei no rekishi to keikan*, vol. 2, Uji-shi shi (Kyoto: Uji Shiyakusho, 1974), pp. 158–160.

40 Fujii Manabu and Moriya Shigeru, "Shūkyō to hōjō shisō," pp. 158–160.

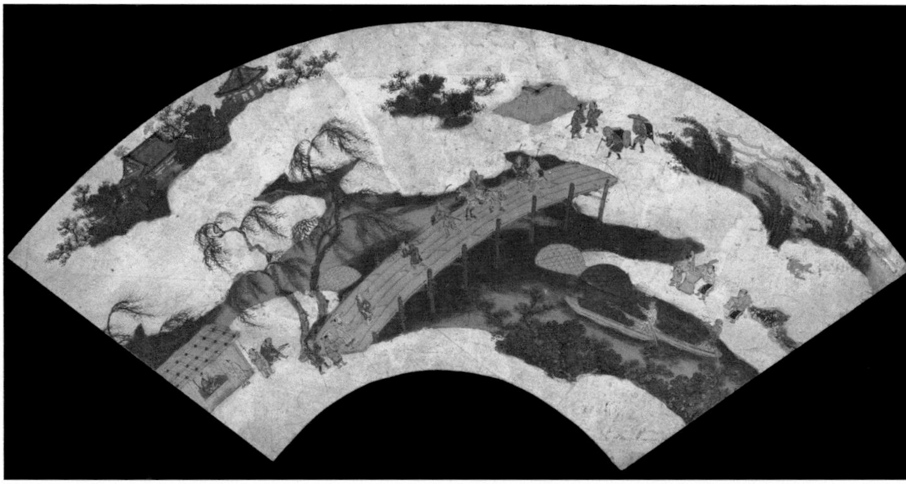

Figure 2.5: Kanō Shōei, *Uji Bridge from Famous Places in the Capital and Legendary Figures*. 16th century. Fan painting, ink, colour, and gold on paper. 19.2cm. Museum of Fine Arts, Boston. William Sturgis Bigelow Collection. 11.6380.8. Photograph ©Museum of Fine Arts, Boston.

Harvesting Tea as Harmonious Labour

The cotton bleaching scene is again closed off with a pine tree; the tea harvesting scene is next – another representation of harmonious labour (Figure 2.6). In the final section of the scroll, Tansetsu shows the interior of a large house where the leaves are piled up in baskets, and tea jars are prepared for storage (Figure 2.7). It looks as if the tea producers have already concluded the whole steaming process and are about to put the final product into jars. Another pine tree ends the scroll on an auspicious note. Throughout the scene, some of the men are relaxing, while the women are diligently picking leaves, with children helping them; the pickers' attire is again too beautiful for farm work. Young and old, men and women, everyone has a share in the scroll.

Uji is famous for its tea production, most notably powdered and whipped tea (*matcha*). Scenes of tea shrubs covered with bamboo mats were iconic for the Uji tea industry as early as the late sixteenth or the early seventeenth century. João Rodrigues (ca. 1562–1633), a Portuguese Jesuit priest and interpreter who stayed in Japan from 1577 to 1610, wrote about the mats spread over the shrubs, a unique method developed in Uji:

> Its new leaves, which are used in the drink, are extremely soft, tender, and delicate, and frost may easily make them wither away. So much damage can be done in this way that in the town of Uji, where the best tea is produced, all the plantations and fields in which this *cha* is grown are covered over with awnings or mats made of

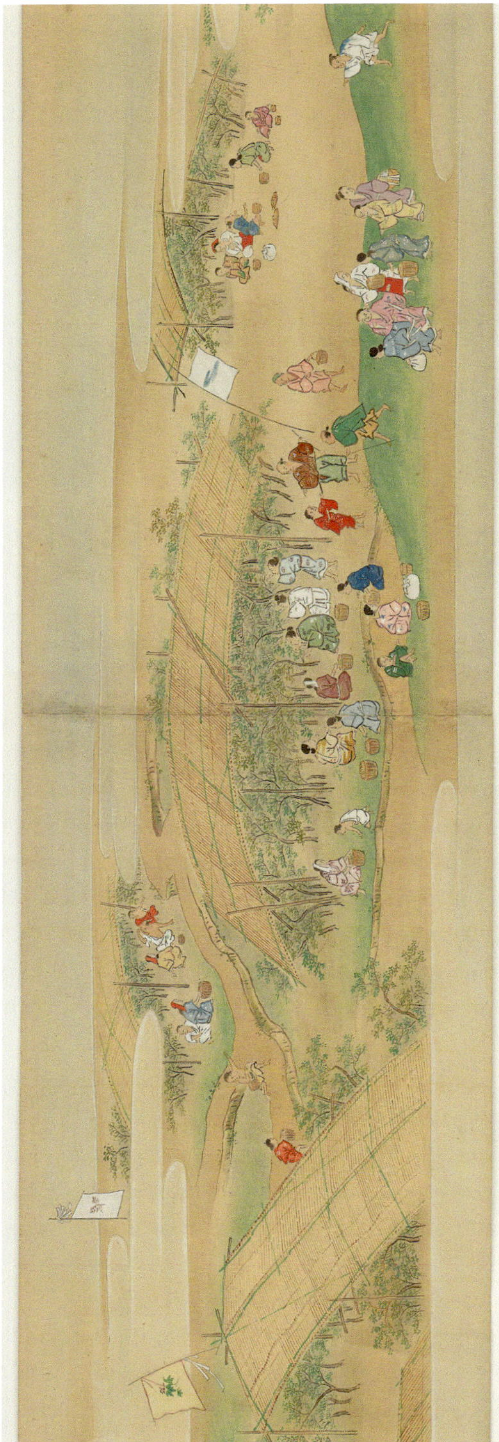

Figure 2.6: Kanō Tansetsu Morisada, *Tea Harvesting at Uji*. 18th century. Chester Beatty Library, CBL J 1137. Part of the scroll pictured. ©The Trustees of the Chester Beatty Library, Dublin.

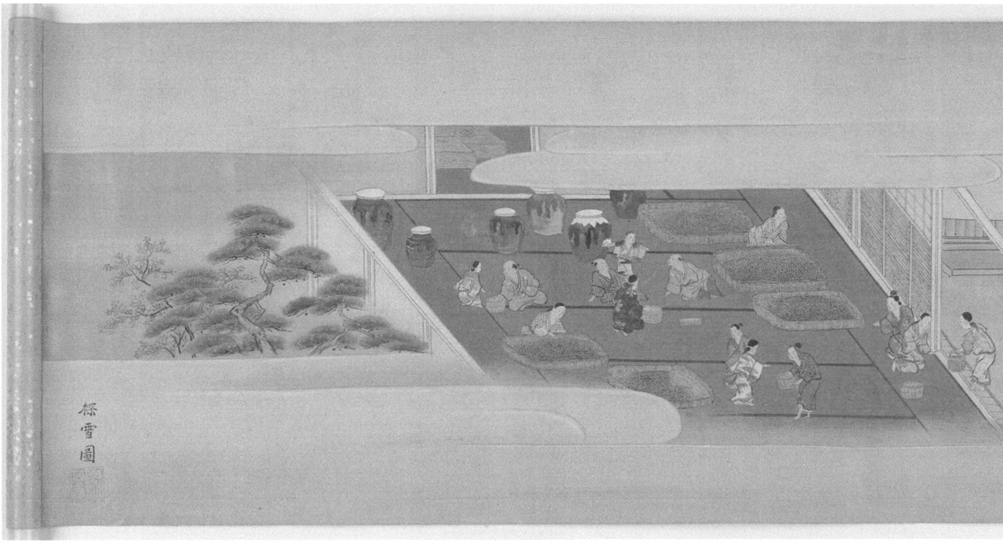

Figure 2.7: Kanō Tansetsu Morisada, *Tea Harvesting at Uji*. 18th century. Chester Beatty Library, CBL J 1137. Part of the scroll pictured. ©The Trustees of the Chester Beatty Library, Dublin.

> rice straw or thatch. They are thus protected from damage by frost from February onwards until the end of March, when the new leaf begins to bud. They spend a great deal of money on this for the sake of the profit that is to be obtained, as we shall say, for the trade in *cha* is very great.[41]

The reason for spreading mats over the shrubs, however, was not to protect the tea leaves from frost, but rather to block the sunlight and to prevent theanine from transforming into bitter-tasting catechin. This process makes the tea taste sweeter.[42] This careful and inventive procedure caught the attention of prominent cultural and political figures of the day, among them tea master Sen no Rikyū, who established tea gatherings in the rustic style known as *wabi cha*, and Toyotomi Hideyoshi, mentioned above, whom Rikyū served. In 1591, Hideyoshi visited Uji to see the harvest on the fifteenth day of the third month.[43] He favoured the newly emerging tea producer Kanbayashi Hisashige (dates unknown) to ensure the quality of his tea and encouraged him to work even harder.[44]

41 João Rodrigues, *João Rodrigues's Account of Sixteenth-Century Japan*, Works Issued by the Hakluyt Society 3 (London: Hakluyt Society, 2001), pp. 272–273.

42 Ōmori Masashi, *Ocha no kagaku: iro, kaori, aji o umidasu chaba no himitsu* (Tokyo: Kōdansha, 2017), p. 30.

43 Uji-shi Rekishi Shiryōkan, *Uji cha: meisho e kara seicha zu e*, p. 24. Tokyo Daigaku Shiryō Hensansho, ed., *Tokitsune-kyō ki*, vol. 4, Dai Nihon kokiroku (Tokyo: Iwanami Shoten, 1964), p. 201.

44 Uji-shi Rekishi Shiryōkan, *Uji cha: meisho e kara seicha zu e*, p. 26.

In the following years, the relationship between Uji's tea production and politics intensified. In 1600, Tokugawa Ieyasu (1542–1616) won the decisive Battle of Sekigahara, which enabled him to seize power. Ieyasu was inaugurated as shogun in 1603 but handed the title to his son Hidetada (1579–1632) in 1605 to establish the Tokugawa family line. The new Tokugawa shogunate included Uji in its direct holdings, and the Kanbayashi family, renowned tea producers and local district intendants since Toyotomi Hideyoshi's time, continued their duties under Tokugawa rule. The Tokugawa shogunate used prominent local figures to protect the tea industry, which had come to play a vital role in samurai culture. A key moment in this respect was the first Procession of the Tea Jars (*ochatsubo dōchū*), which took place in 1633 during the reign of the third shogun, Tokugawa Iemitsu (1604–1651). The tea produced by the Kanbayashi family for the shogun and his close associates was carried in tea jars to the shogunal capital Edo. Officials were dispatched from Edo to Uji between the fourth and fifth month to supervise the packing of the tea into the jars and their transportation. The way back to Edo was carefully guarded by officials and members of local households, which testifies to the prestige of the product and to the power of the shogunate.[45] The procession continued throughout the Edo period.

By the late seventeenth century, tea produced in Uji became widely known as the product that represented the region. *The Illustrated Encyclopedia of Humanity* lists tea producers at the top of the section on primary sectors of agriculture and adds the following information to an image of a tea harvest:

> Tea producers. Tea was brought from China. The founder of the Kennin-ji Temple, Yōsai, went to the country of Song and came back to Japan with the seeds of tea. Also, Abbot Myōe planted tea on Mt. Togano'o. It is said that this was the beginning [of tea production]. The legend goes that the monks of this mountain made their living from producing tea until the Middle Ages. Subsequently, tea was widely planted in the area around Uji, making this [drinking of tea] into a beautiful and enjoyable custom. The manners of tea ceremony have been set and transmitted to our times. As for the steeped tea, it is available everywhere because all the cities and countryside produce it.[46]

In the image, a tea master supervises the harvest in the Kanbayashi field. The family name is given on the flag, and from it we can infer that the area is Uji. The stern facial expression of the tea master emphasises strict supervision and quality

45 For details, Taka Oshikiri, "The Shogun's Tea Jar: Ritual, Material Culture, and Political Authority in Early Modern Japan," *The Historical Journal* 59, no. 4 (2016): pp. 927–945.

46 Asakura Haruhiko, ed., *Jinrin kinmō zui*, p. 86. For the illustration, see the digital collection in the National Diet Library, Tokyo (https://dl.ndl.go.jp/pid/2592441/1/4).

control. Women of all ages are engaged in picking tea leaves and collecting them in baskets. A man with a rope basket on his shoulder may be carrying lunch for the pickers. Together, text and image provide information on the locality of production, the availability of the products, and the importance of tea producers in Uji.[47]

As the text says, the tea ceremony was a major factor in perpetuating the fame of Uji's tea. The population of tea practitioners grew rapidly in the Genroku period (1688–1703), a time when society became more peaceful and stable. The year 1690 was counted as the hundredth anniversary of Sen no Rikyū's death, and the return to Rikyū's *wabi* style tea was greatly promoted.[48] Printed books on tea practice were published in huge numbers, assisting novice practitioners in learning the tea ceremony.[49] Many wealthy townspeople began taking up hobbies, including the tea ceremony, to create their networks, hoping to use it to make connections in towns full of strangers and newcomers.[50] Such was the increase of information on the tea ceremony and on tea that even Engelbert Kaempfer (1651–1716), a German doctor who came to Japan in 1690, described Uji and its tea in his diary.[51] The different ways in which tea was explained in Nakagawa Kiun's (d. 1705) *Denizens of Kyoto* (Kyō warabe, published in 1658) and in *The Illustrated Encyclopedia of Humanity,* published thirty-two years later, demonstrate the increasing social and cultural status tea held in the late seventeenth century. The former explained tea as a valuable thing helping people to mingle, stay awake, and clear their chest while the latter described it as "a beautiful and enjoyable custom," emphasising the establishment of the tea ceremony. It is fair to say, then, that tea had become widely known as a *famous product* (*meibutsu*) of Uji and that it had become deeply embedded in urban socialisation culture by the late seventeenth century.

47 It is possible that one of the reasons why the Kanbayashi tea producers are listed at the top in this section was their role of chief magistrate serving the shogunate. A change in the order of listing things from the formal analysis to the social status of the owner of the object is also seen in the list of famous tea objects since *Ganka meibutsu ki* published in 1660, in which these were listed beginning from the shogun, the Three Collateral Houses of the Tokugawa and followed by the other daimyo. See Takeuchi Jun'ichi, "Meibutsu ki no seisei kōzō: jikken to henshū no hazama dai ichi bu," in *Cha no koten*, ed. Tsutsui Hiroichi, vol. 10, Chadōgaku taikei (Kyoto: Tankōsha, 2001), pp. 45–62.

48 Paul Varley, "Chanoyu: From the Genroku Epoch to Modern Times," in *Tea in Japan: Essays on the History of Chanoyu*, ed. Paul Varley and Kumakura Isao (Honolulu: University of Hawaii Press, 1989), 165–168.

49 Murai Yasuhiko, "Kinsei chadō no tenkai," in *Chanoyu no tenkai,* vol. 5, Chadō shūkin (Tokyo: Shōgakukan, 1985), pp. 43–44.

50 Moriya Takeshi, *Genroku bunka: yūgei, akusho, shibai,* Nippon sōshi (Tokyo: Kōbundō, 1987), pp. 38–41. Eiko Ikegami, *Bonds of Civility: Aesthetic Networks and the Political Origins of Japanese Culture* (Cambridge: Cambridge University Press, 2005), pp. 140–144.

51 Engelbert Kaempfer, *Kaempfer's Japan: Tokugawa Culture Observed,* ed. and trans. Beatrice M Bodart-Bailey (Honolulu: University of Hawaii Press, 1999), p. 318.

This context may explain why Tansetsu's scroll was commissioned. Picturing famous agricultural products and their area of production may have been driven by the growing fame of the product itself. Three points should be made in this regard, however. Firstly, tea harvesting at Uji was probably one of the earliest themes linking a particular type of agricultural product with a specific area of production. In addition to the social and cultural importance that tea gained, as discussed above, a change in the idea of natural resources occurred, from sacred things to natural products used for growing wealth and for human welfare.[52] This may have encouraged people to focus on the product and the process of its making. However, not all products were depicted in paintings. What made tea harvesting at Uji a suitable theme for a scroll painting?

Secondly, the fact that Uji was known as a famous poetic and religious place, and the fact that a framework existed for depicting it as such, provided opportunities for depicting its tea industry in painting.[53] The question of what gets to be represented and what is left out is inherently difficult to answer in the history of images, but it is fair to say that being a *famous place* helped the representation of Uji's tea production along. Considering this, should we assume that Tansetsu's scroll simply integrated tea harvesting scenes into the existing image of Uji as a *famous place*? The fact that some iconic features are missing, and the depiction of the waterwheel as a practical tool suggests otherwise. Tansetsu's scroll builds on existing pictorial traditions, but its focus is on people at work.[54] Similarly, the theme of tilling and weaving falls too short to serve as the framework for Tansetsu's scroll. As we have seen, it is difficult to sense admonitory references; in addition, he selected a few highlights, not the whole process of cultivation. Rather, the depicted industries form a backdrop for showcasing the figures: their gestures and eye contact drive the

52 Art historian Christine Guth discusses this point in the first chapter of Christine Guth, *Craft Culture in Early Modern Japan: Materials, Makers, and Mastery*, University of Kansas Franklin D. Murphy Lecture Series (Oakland, CA: University of California Press, 2021). As for the natural resources' role as economic revenue sought by the eighth shogun Tokugawa Yoshimune, see Federico Marcon, *The Knowledge of Nature and the Nature of Knowledge in Early Modern Japan*, Studies of the Weatherhead East Asian Institute (Chicago, IL: University of Chicago Press, 2015), pp. 115–124.

53 A scroll depicting cotton production in what might be the Kawachi region (present-day Osaka), which was famous for the industry, is in the collection of the Asian Art Museum, National Museums of Berlin. It is argued that this scroll follows the traditional pictorial language of the image of weaving in the *Weaving and Tilling*, but the place in the image is not clearly identifiable as Kawachi, despite what the current title suggests. See Nagai Ken, "'Kawachi momen seishoku zukan' o megutte," in *Bi ga musubu kizuna: Berurin Kokuritsu Ajia Bijutsukan shozō Nihon bijutsu meihin ten,* ed. Kōriyama Ichiritsu Bijutsukan et al. (Tokyo: Howaito Intānashonaru, 2008), pp. 25–26.

54 In this regard, Tansetsu's continuation of his father's legacy should be evaluated in relation to Kusumi Morikage and Hanabusa Itchō (1652–1724), two former Kanō artists who were expelled by their master (Morikage was by Tan'yū and Itchō was by Yasunobu, Tan'yū's youngest brother) but are regarded highly as painters of people's daily lives.

viewer's gaze and the pictorial narrative of the scroll. By 'populating' Uji, Tansetsu creates a new, human-centred image of Uji, thereby challenging the poetical and religious biases of earlier imagery.

Lastly, then, why did the image of tea cultivation emerge in the late seventeenth century, a rather late birth, considering that it was already renowned by the beginning of the seventeenth century? This late emergence may explain the length of time required for a new image to be accepted by society. Images of tea harvesting are rarely seen until the mid-seventeenth century. One of the earliest images can be found in the fictional travel gazetteer *A Kyoto Souvenir from Dekisai*, published in 1677. The book features an imaginary traveller named Dekisai who travels to Kyoto and writes about what he has seen there. Tea harvesting in Uji is depicted together with text describing how people in Uji delicately tend to their shrubs.[55] Gazetteers of this kind and guidebooks to *famous places* were part of the travelling craze that emerged in the late seventeenth century. *Tea Harvesting at Uji* by Tansetsu emerged at the nexus of these social, cultural, and iconological conditions.

Re-framing the Ideal

Regarding travel, was the wish to travel also a motivation behind the production of Tansetsu's scroll? It may have been, but elements such as the deliberate change of the current of Uji River and the replacement of the Hashihime Shrine with an ordinary house suggest otherwise, even if the scroll was only intended for the 'armchair traveller.' In the final scene of Tansetsu's scroll, we see a child dancing with a fan in front of tea jars filled with tea leaves, which adds to the cheerful and celebratory atmosphere, suggesting that there was no need to worry about the safe delivery of these jars. These highlights sufficed as carriers of the idea of Uji as a place where diligent people produced various products including tea.

The purpose of Tansetsu's scroll was not to depict reality. Its objective was to create a fictional space. From beginning to end, the images are framed by two streams of mist at the top and bottom. This gives the viewer the impression that they are peering into a space that is set apart from their own. The pine trees, symbols of eternity used to mark the scroll's various sections, can be seen as another frame; the division into sections underlines the scenes' fictional character.[56] These horizontal and vertical frames suggest a glimpse into an ideal world, in which people dressed

55 Asai Ryōi, "Dekisai Kyō miyage," pp. 649–650. For the illustration, see the digital collection of the Ranpo Bunko, Rikkyō University provided by the National Institute of Japanese Literature (https://kotenseki.nijl.ac.jp/api/iiif/100289203/v4/RKRP/RKRP-00053/RKRP-00053-0212.tif/full/full/0/default.jpg).

56 The role of rocks and pine trees as the frames that indicate the ideal character of the image is discussed by Satō Yasuhiro, "Takao kanpū zu ron," *Bijutsushi ronsō*, no. 16 (1999): pp. 25–26.

in finely decorated attire interact harmoniously, in the best season for harvesting and for packing the product.

These frames were effective tools for introducing a new image of a particular place. Frames made viewers conscious of the fact that they were looking at an image that was separate from the world they were in. This is clearly seen at the beginning of the scroll, where locals standing on Uji Bridge demonstrably look at travellers as if to differentiate themselves from the casual visitors. At the same time, the locals are also objects of the gaze of viewers; the frame presents and emphasises the 'daily' aspects of their life. From a different point of view, this framing might also have prevented viewers from considering that productivity was not necessarily the same as prosperity. Although the yearly tax levied on tea masters was reduced to half the initial amount in 1636, two years after Iemitsu's procession to Kyoto, its economic impact did not last long. Although commodity prices went up, the price of tea remained fixed since 1642. In addition, frequent fires (1670, 1684, 1697, 1696) and flooding damaged the industry, resulting in the accumulation of debt by tea producers.[57]

Conclusion

There is a tautology in the meaning of *famous place*; it is a place that is famous for something. This 'something' changes with time and according to people's inclinations. In addition, the image of a place changes if accepted by most people. Uji as a poetic and religious place was the product of the imagination shared by people who largely belonged to the upper strata of society before and well into the early modern period. These meanings did not disappear completely, of course, but the image of Uji as a tea production site gained a wider appeal among more diverse audiences in the late seventeenth century. This was driven by social, cultural, and historical changes affecting Uji as well as the culture, production, and the consumption of tea.

Kanō Tansetsu's *Tea Harvesting at Uji* emerged at the nexus of these changes. Tansetsu dismantled the traditional image of a religious and poetic Uji and reassembled the fragments to show a human-centred Uji. The activities he depicted – storing rice, bleaching cotton, and harvesting tea leaves – were the highlights of industries requiring human hands. In this way, Tansetsu highlighted cooperative labour and its productivity. The scroll shows an ideal world framed by mist and pines that underscore the scenes' imaginary character. In quite a revolutionary manner, Tansetsu used simple human activities and pleasures to present a novel way of imagining a place that had once been fraught with pessimism and sadness.

57 Wakahara Eiichi and Yoshimura Tōru, "Chashi nakama to chatsubo dōchū," in *Kinsei no rekishi to keikan*, vol. 3, Uji-shi shi (Kyoto: Uji Shiyakusho, 1976), pp. 156–161.

Glossary

Asai Ryōi	浅井了意
Amida Hall	阿弥陀堂
Baika mujinzō	梅花無尽蔵
Banri Shūku (1428–?)	万里 集九
Byōdō-in	平等院
Chūson-ji	中尊寺
Dekisai Kyō miyage	出来斎京土産
Eison (1201–1290)	叡尊
Eiri Genji monogatari	絵入源氏物語
Fujiwara no Yorimichi (992–1074)	藤原頼通
Fushimi Castle	伏見城
Ganka meibutsu ki	玩貨名物記
Genroku	元禄
hasami bako	挟箱
Hashihime Shrine	橋姫神社
Hiraizumi	平泉
Jinrin kinmō zui	人倫訓蒙図彙
Kaibara Ekiken (1630 –1714)	貝原益軒
Kaihō Yūsen (1690–1741)	海北友泉
Kanbayashi Hisashige (? – 1589)	上林久茂
Kanjizaiō-in	観自在王院
Kanō Eitoku (1543–1590)	狩野永徳
Kanō Shōei (1519–1592)	狩野松栄
Kanō Tansetsu (1655–1714)	狩野探雪
Kanō Tan'yū (1602–74)	狩野探幽
Katsura River	桂川
Kenkō (1283–1350)	兼好
Keijō shōran	京城勝覽
Kisen (dates unknown)	喜撰
Kizu River	木津川
Kokin wakashū	古今和歌集
Kusumi Morikage (dates unknown)	久隅守景
Kyō warabe	京童
Kyahan	脚絆
Makinoshima	槙島
Manpuku-ji	萬福寺
Masu	枡
Matcha	抹茶

Matsunaga Hisahide (1510–1577)	松永久秀
Meibutsu	名物
Meisho	名所
Miyako meisho zue	都名所図会
Nakagawa Kiun (d. 1705)	中川喜雲
Nara	奈良
Ōbaku Zen	黄檗宗
Oda Nobunaga (1534–1582)	織田信長
Ochatsubo dōchū	御茶壷道中
Rakushidō nenroku	樂只堂年錄
Sadayori shū	定頼集
Saidai-ji	西大寺
san-no-ma	三の間
Sekigahara	関ヶ原
Sen no Rikyū (1522–1591)	千利休
Song Zonglu (dates unknown)	宋宗魯
Tokugawa Ieyasu (1542–1616)	徳川家康
Tokugawa Iemitsu (1604–1651)	徳川家光
Tokugawa Tsunayoshi (1646–1709)	徳川 綱吉
Toyotomi Hideyoshi (1536–1598)	豊臣秀吉
Tsurezuregusa	徒然草
Uji	宇治
Utamakura	歌枕
wabi	侘
Waka	和歌
Yamamoto Shunshō (1610–1682)	山本春正
Yamashiro	山城
Yanagisawa Yoshiyasu (1658–1714)	柳沢吉保
Yinyuan Longqi (Jap. Ingen Ryūki, 1592–1673)	隱元隆琦
Yodo River	淀川

Bibliography

Primary Sources

Asai Ryōi 浅井了意. "Dekisai Kyō miyage 出来斎京土産," in Noma Kōshin 野間光辰 ed., *Shinshū Kyōto sōsho* 新修京都叢書, 2nd ed., 11:487–658. Kyoto. Rinsen Shoten, 1976.

Asakura Haruhiko 朝倉治彦, ed. *Jinrin kinmō zui* 人倫訓蒙図彙. Tōyō bunko 東洋文庫 519. Heibonsha, 1990.

Miyakawa Yōko 宮川葉子, ed. *Rakushidō nenroku* 楽只堂年録. Vol. 1. Tokyo. Yagi Shoten, 2011.

Nakagawa Kiun 中川喜雲. "Kyō warabe 京童," in Noma Kōshin 野間光辰 ed., *Shinshū Kyoto sōsho* 新修京都叢書, 2nd ed., 1:1–88. Kyoto. Rinsen Shoten, 1976.

Tokyo Daigaku Shiryō Hensansho 東京大学史料編纂所, ed. *Tokitsune-kyō ki* 言経卿記. Vol. 4. Dai Nihon kokiroku 大日本古記録. Tokyo. Iwanami Shoten, 1964.

Yoshida Kenkō 吉田兼好. "Tsurezuregusa 徒然草," in *Shinpen Nihon koten bungaku zenshū*, translated by Nagazumi Yasuaki 永積安明, 44:67–309. Tokyo. Shōgakukan, 1995.

Secondary Sources

Adachi Keiko 安達啓子. "Nishu no ryūkyō zu byōbu 二種の柳橋図屏風." *Kokka* 国華, no. 1135 (1990): 3–22.

Ashikaga Kenryō 足利健亮, Mizuyama Takayuki 水山高幸, Akiyama Motohide 秋山元秀, Takahashi Seiichi 高橋誠一, and Wakahara Eiichi 若原英弌. "Uji-gawa no chisui 宇治川の治水," in *Chūsei no rekishi to keikan* 中世の歴史と景観, 2:562–625. Uji-shi shi 宇治市史. Kyoto: Uji Shiyakusho 宇治市役所, 1974.

Fujii Manabu 藤井学 and Moriya Shigeru 守屋茂. "Shūkyō to hōjō shisō 宗教と放生思想," in *Chūsei no rekishi to keikan* 中世の歴史と景観, 2:128–62. Uji-shi shi 宇治市史. Kyoto: Uji Shiyakusho 宇治市役所, 1974.

Gerhart, Karen M. "Talent, Training, and Power: The Kano Painting Workshop in the Seventeenth Century," in Brenda G. Jordan and Victoria Weston eds., *Copying the Master and Stealing His Secrets: Talent and Training in Japanese Painting.* Honolulu: University of Hawai'i Press, 2003, pp. 9–30.

Goree, Robert. *Printing Landmarks: Popular Geography and Meisho Zue in Late Tokugawa Japan*. Cambridge, MA: Harvard University Asia Center, 2020.

Guth, Christine. *Craft Culture in Early Modern Japan: Materials, Makers, and Mastery.* Oakland, CA: University of California Press, 2021.

Hammers, Roslyn Lee. *Pictures of Tilling and Weaving: Art, Labor and Technology in Song and Yuan China*. Hong Kong: Hong Kong University Press, 2011.

Hirayama Ikuo 平山郁夫, Kobayashi Tadashi 小林忠, eds. *Chestā Bītiī Raiburarī* チェスター ビーティ ライブラリー. Vol. 5. Hizō Nihon bijutsu taikan 秘蔵日本美術大観. Tokyo: Kōdansha, 1996.

Ichiki Takeo 市木武雄, trans. *Baika mujinzō chūshaku* 梅花無尽蔵注釈. Vol. 1. Tokyo: Zoku Gunsho Ruijū Kanseikai, 1993.

Ikegami, Eiko. *Bonds of Civility: Aesthetic Networks and the Political Origins of Japanese Culture.* Cambridge: Cambridge University Press, 2005.

Jordan, Brenda G. "Copying from Beginning to End? Student Life in the Kano School," in Brenda G. Jordan and Victoria Weston eds., *Copying the Master and Stealing His Secrets: Talent and Training in Japanese Painting.* Honolulu: University of Hawai'i Press, 2003, pp. 31–59.

Kaempfer, Engelbert. *Kaempfer's Japan: Tokugawa Culture Observed.* Edited and translated by Beatrice M. Bodart-Bailey. Honolulu: University of Hawaii Press, 1999.

Kamens, Edward. *Utamakura, Allusion, and Intertextuality in Traditional Japanese Poetry.* New Haven, CT: Yale University Press, 1997.

Kano Hiroyuki 狩野博幸. "Kanō Eitoku hitsu Rakugai meisho yūraku zu byōbu (Ujigawa, Ōigawa) 狩野永徳筆 洛外名所遊樂圖屏風 (宇治川・大堰川)." *Kokka* 國華, no. 1331 (2006): 19–22.

Marcon, Federico. *The Knowledge of Nature and the Nature of Knowledge in Early Modern Japan*. Chicago, IL: University of Chicago Press, 2015.

McKelway, Matthew P. *Capitalscapes: Folding Screens and Political Imagination in Late Medieval Kyoto*. Honolulu: University of Hawai'i Press, 2006.

Moriya Takeshi 守屋毅. *Genroku bunka: yūgei, akusho, shibai* 元禄文化 遊芸・悪所・芝居. Nippon sōshi にっぽん草子. Tokyo: Kōbundō, 1987.

Murai Yasuhiko 村井康彦. "Kinsei chadō no tenkai 近世茶道の展開," in *Chanoyu no tenkai* 茶の湯の展開, 5:41–51. Chadō shūkin 茶道聚錦. Tokyo: Shōgakukan, 1985.

Nagai Ken 長井健. "'Kawachi momen seishoku zukan' o megutte「河内木綿製織図巻」をめぐって," in Kōriyama Ichiritsu Bijutsukan 郡山市立美術館, Iwate Kenritsu Bijutsukan 岩手県立美術館, Yamaguchi Kenritsu Bijutsukan 山口県立美術館, and Ehime Kenritsu Bijutsukan 愛媛県立美術館 eds., *Bi ga musubu kizuna. Berurin Kokuritsu Ajia Bijutsukan shozō Nihon bijutsu meihin ten* 美がむすぶ絆 ベルリン国立アジア美術館所蔵日本美術名品展. Tokyo: Howaito Intānashonaru, 2008, pp. 24–27.

Noguchi Takeshi 野口剛. "Uji zu byōbu 宇治図屏風." *Kokka* 国華, no. 1392 (2011): 24–28.

Ōmori Masashi. *Ocha no kagaku: iro, kaori, aji o umidasu chaba no himitsu* お茶の科学 色・香・味を生み出す茶葉のひみつ. Tokyo: Kōdansha, 2017.

Oshikiri, Taka. "The Shogun's Tea Jar: Ritual, Material Culture, and Political Authority in Early Modern Japan." *The Historical Journal* 59, no. 4 (2016): 927–945.

Reizei Tamehito 冷泉為人, Kōno Michiaki 河野通明, Iwasaki Takehiko 岩崎竹彦, and Namiki Seishi 並木誠士. *Mizuho no kuni Nihon. shiki kōsaku zu no sekai* 瑞穂の国日本 四季耕作図の世界. Kyoto: Tankōsha, 1996.

Rodrigues, João. *João Rodrigues's Account of Sixteenth-Century Japan.* Works Issued by the Hakluyt Society 3. London: Hakluyt Society, 2001.

Satō Yasuhiro 佐藤康宏. "Takao kanpū zu ron 高雄観楓図論," *Bijutsushi ronsō* 美術史論叢, no. 16 (1999). 1–48.

Screech, Timon. *Obtaining Images: Art, Production and Display in Edo Japan*. London: Reaktion Books, 2012.

Takeda Tsuneo 武田恒夫, Murashige Yasushi 村重寧, Adachi Keiko 安達啓子, Inoue Ken'ichirō 井上研一郎, and Yoshida Tomoyuki 吉田友之. *Keibutsu ga: shiki keibutsu* 景物画 四季景物. Edited by Daiichi Shuppan Sentā 第一出版センター. Vol. 9. Nihon byōbu e shūsei 日本屏風絵集成. Tokyo: Kōdansha 講談社, 1977.

Takeuchi Jun'ichi 竹内順一. "Meibutsu ki no seisei kōzō: jikken to henshū no hazama dai ichi bu 名物記の生成構造 実見と編集のはざま 第一部," in Tsutsui Hiroichi 筒

井紘一 ed., *Cha no koten* 茶の古典, 10:45–62. Chadōgaku taikei 茶道学大系. Kyoto: Tankōsha, 2001.

Takeuchi Misako 竹内美砂子. "Ryūkyō suisha zu byōbu: shinshutsu bon no shōkai o kanete jō 柳橋水車図屏風 新出本の紹介をかねて 上." *Kokka* 国華, no. 1138 (1990): 20–34.

Tamamushi Satoko 玉蟲敏子. "'Ryūkyō suisha zu' to 'Uji no kawase no mizuguruma': kayō to fūryū to byōbu e no kōkyō 「柳橋水車図」と「宇治の川瀬の水車」歌謡と風流と屏風絵の交響," in *Nihon bungei shichōron* 日本文芸思潮論. Tokyo: Ōfūsha, 1991, pp. 859–880.

Tsuji Nobuo 辻惟雄. *Rakuchū rakugai zu* 洛中洛外図. Vol. 121. Nihon no bijutsu 日本の美術. Tokyo: Shibundō, 1976.

Uji-shi Rekishi Shiryōkan 宇治市歴史資料館. *Uji-bashi: sono rekishi to bi to* 宇治橋 その歴史と美と. Uji 宇治: Uji-shi Rekishi Shiryōkan, 1995.

Uji-shi Rekishi Shiryōkan 宇治市歴史資料館. *Uji cha: meisho e kara seicha zu e* 宇治茶名所絵から製茶図へ. Uji 宇治: Uji-shi Rekishi Shiryōkan, 1985.

Varley, Paul. "Chanoyu: From the Genroku Epoch to Modern Times," in Paul Varley and Kumakura Isao eds., *Tea in Japan: Essays on the History of Chanoyu*. Honolulu: University of Hawaii Press, 1989, pp. 161–194.

Wakahara Eiichi 若原英弌 and Yoshimura Tōru 吉村亨. "Chashi nakama to chatsubo dōchū 茶師仲ヶ間と茶壺道中," in *Kinsei no rekishi to keikan* 近世の歴史と景観, 3:129–196. Uji-shi shi 宇治市史. Kyoto: Uji Shiyakusho, 1976.

About the Author

Shiori Hiraki is an art historian working on art in social life in the early modern period. She was awarded a PhD degree from SOAS, University of London, in 2021. Her thesis, *Onari: Art, Ritual and Power in Early Modern Japan*, focuses on the use of art in *onari*, the Tokugawa shogunal visit to the daimyo residences in Edo. From paintings to ceramics and lacquers to gardens, she has argued that these objects were articulated to create a special setting to welcome the shogun and to represent his identity as the top of the military government. She was awarded a postdoctoral fellowship from the Reischauer Institute of Japanese Studies, Harvard University (2021–2022), and has taught at School of International Liberal Studies Waseda University. She is currently turning her doctoral thesis into a book manuscript that will focus on the shogunal visits by the fifth shogun, Tokugawa Tsunayoshi.

3. Disciplined Objects? Wood panels from the Kew Collections[1]

Maki Fukuoka

Abstract: This chapter analyses a set of twenty-six wooden panels currently held at Kew's Economic Botany Collection. Katō Chikusai produced the set at Koishikawa Botanical Garden in Tokyo in 1878. By taking seriously the methodological and discursive presuppositions that are often and easily overlooked in studies of early Meiji visual objects, this chapter reorients the perspective to one that privileges the context of the production site of the wood panels over their current location. Through historically informed analyses, this chapter highlights how these objects carry two constitutive values – economic and epistemic – of early Meiji Japan.

Keywords: Knowledge production, material culture, Meiji, botanical garden, Katō Chikusai

The central object of analysis in this chapter is a set of twenty-six wood panels currently held at Kew's Economic Botany Collection.[2] The set was produced in

1 The initial research of this chapter's content began in 2011 when Christine Guth shared with me her encounter with these objects. I then presented a version of my analysis at "Curious and Scientific Things: Seeing and Knowing in Early Modern East Asia," organised by Doreen Mueller and Fan Lin at Leiden University in October 2018, where I received constructive feedback. Further conversations with Kurata Aiko and Richard Checketts were instrumental in the following research and thinking. I am grateful for their support and feedback. I also want to acknowledge Doreen and Fan's tireless effort to bring this publication forward and the support from proofreaders we received throughout the process.

2 A larger collection consisting of 152 panels, also bearing the name of Katō Chikusai, is at the Botanisches Museum Berlin-Dahlem. They are said to have originally belonged to the Schaumuseum of the Botanisches Museum Berlin-Dahlem. Moreover, eight comparable panels are located in the Harvard University Herbaria, and nine are in a private collection in London. Recently, another twenty-five panels were discovered at the Koishikawa Botanical Garden. Given the time and space constraints of the present volume, this chapter focuses on the Kew panels. For the most recent comparative investigation of these sets, see Toshiyuki Nagata, et. al., "An Unusual Xylotheque with Plant Illustrations from Early Meiji Japan," in *Economic Botany* 67(2) and H. Walter Lack, "Plant Illustration on Wood Blocks: A Magnificent

Fan Lin and Doreen Mueller (eds), *Picturing Animals and Plants in Early Modern China and Japan. Innovation, Experiments, and Anxieties.* Taylor & Francis Group, 2025
DOI: 10.5117/9789048559091_CH03

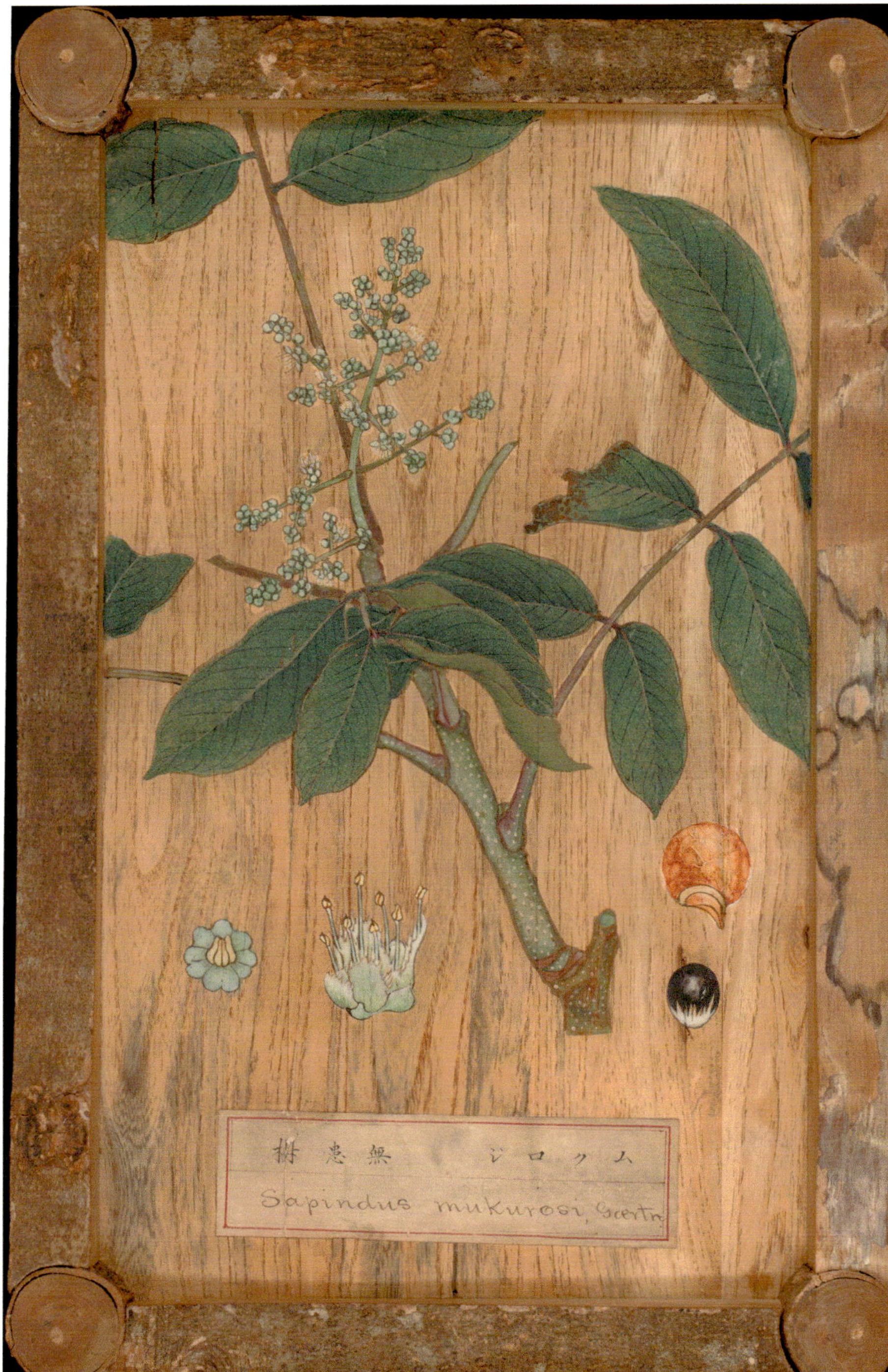

Figure 3.1: Katō Chikusai, *Sapindus Mukorossi (Mukuroji)*, 1878. Tempera on wood. Kew Gardens, London.

1878 at the Koishikawa Botanical Garden in Tokyo. Each of the twenty-six panels bears a stamp with Katō Chikusai (1818–1886)'s name on the reverse in red ink.[3] Measuring about 34 x 23 x 3 cm, each panel uses at least nine physical pieces from an actual tree: 1) one rectangular board cut flat-sawn from the trunk of the tree; 2) four corner pieces made from the branches of the tree cut quartersawn; and 3) four rectangular pieces of the bark and sapwood placed on the edges of the central board. These pieces have been fitted together with glue and nails. The front of the central section carries a pictorial representation of the tip of branches of the tree in question, including flowers and leaves, and fruit if applicable, in what appears to be tempera pigment (Figure 3.1). A white label is attached to the back. It includes the name of the tree in Chinese characters and Japanese *katakana* script, as well as a Latin binominal name in Roman alphabet. How these panels came to the Kew is unknown.[4]

In the recent discussions of these wood panels, scholars underscore the shared similarities with what is referred to as a 'xylotheque.'[5] This term refers to a collection of wood samples and its history dates to late-seventeenth-century Europe, mainly found in Germany. The standard form of a specimen in a xylotheque takes the form of a book, its 'spine' covered by the bark of the tree in question, and its contents consisting of dried specimens of leaves, twigs, and fruits. The wood panels at the Kew are posited as an unusual departure from this prototype in that they provide visual representations instead of actual parts. Nonetheless, on a visual level, the similarity of format between a xylotheque and the wood panels as well as the panels' use of Linnaean taxonomy appear to privilege the putatively 'Western' science of botany.[6] These wood panels strike contemporary viewers and scholars alike with a sense of surprise.

Japanese Xylotheque of The Early Meiji Period," in *Curtis' Botanical Magazine,* vol. 16, no.2 (1999). The following paper, in Japanese, provides detailed information and analysis based on recent reading of logs from the Koishikawa Botanical Garden: Murata Hiroko, Sakazaki Nobuyuki, Sakurada Michio, Tōma Tetsuo, Murata Jin, "'Meiji jūshichinen jūnigatsu Tokyo Daigaku shokubutsuen shaseizu (mokuroku)' oyobi Koishikawa Shokubutsuen shōzō mokuzai hengaku nitsuite," in *Itō Keisuke nikki* [The Diary of Itō Keisuke], vol. 19 (Nagoya: Higashiyama Botanical Garden, 2014). I would like to thank Christine E. Guth for letting me know about these objects many years ago.

3 The year of Katō's death has been taken from Nagata Toshiyuki, "Shokubutsugahō sōmeiki no tsuiseki – Katō Chikusai hengaku kara," in *Seibutsu no kagaku iden,* vol. 71, no.3 (May 2017): 191.

4 Mark Nesbitt, "Puzzling Panels," *Kew Magazine* (Spring 2010): 54–55.

5 The term 'xylotheque' consists of the Greek words *xylon* (meaning wood) and *thêkê* (meaning place, and by extension repository). Another term is 'xylarium,' consisting of *xylon* and the Latin word *arium* meaning 'location' or 'receptacle.'

6 For convincing expositions on the concepts of 'the West,' 'the East,' 'the Orient,' and 'the Occident,' see Martin W. Lewis and Kären Wigen, *The Myth of Continents: A Critique of Metageography* (Chicago, IL: University of Chicago Press, 1997), especially "Chapter 2: The Spatial Constructs of Orient and Occident, East and West."

This sense of the unanticipated, I suggest, arises from a particular habitual and often unexamined methodological assumption. In this postulation, the application of Latin nomenclature and the adaptation of the xylotheque seem to exemplify 'Western influence' on these objects, and Japan in the nineteenth century is imagined to be distanced from immediate and explicit sites of *botanical* practice. Indeed, the locations of the collections in which the panels are held, including Kew, intensifies the desire to view the objects as being enveloped by stages of Western influence – first the absorption of Linnaean nomenclature, then the format of the xylotheque. The pictorial rendition on the board appears to cohere with Western pictorial traditions. One notices, for example, that the appearance of the leaves is conveyed through subtle differentiations in hues, resembling contemporary European illustrations of botanical specimens. The prevailing characterisation of the 1868 Meiji Restoration as the beginning of Westernisation of the country heightens the propensity to frame these panels as objects attesting to a socio-political pivot, and the changes in the direction and mechanism for knowledge production that followed.

Peter Dear, a historian of science, asserted that the development of science is a story of two lines of investigations: that of natural philosophy and the other of instrumentality. Dear further asserted that intelligibility is key in making natural philosophy plausible.[7] While Dear was mainly concerned with histories of science in the West, his formula is elucidating in relation to these wood panels because the objects are also marked by particular moments of plant science and art history in Japan.

If we follow his further assertion that "the world pictures that we believe in owe much more to what we find plausible than to the way the world 'really' is: their acceptance, rather than being determined by the natural world itself, depends on the ways in which we choose to live in the world," then this group of objects begs us to examine from which world we are assessing them, to which world we think the objects belong, and how flexible our framing worldview could and should be. I will argue that this group of objects carry layered marks and significances. They are attached to values informed by multiple concerns – epistemic, economic, legal, and pictorial – which emerged both in and outside mid-nineteenth-century Japan.

Disciplinarity and the Wood Panels

Does the periodisation, positing Tokugawa as feudal and Meiji as Westernising periods, account for the ways objects look, or the way they function, or the way

7 Peter Dear, *The Intelligibility of Nature: How Science Makes Sense of the World* (Chicago, IL: The University of Chicago Press, 2006), 194.

they are put together? We know that in the previous Tokugawa period, studies of *materia medica* (Ch: *bencao,* J: *honzō*), which constituted the key component of medical knowledge and practice, took elements of botanical knowledge from publications on natural history in Latin.[8] Moreover, the framework and the utility of this body of knowledge remained embedded within medical knowledge well after the 1868 Meiji Restoration.

It is helpful to situate the often-quoted remark made about the wood panels in 1878 by Edward S. Morse, an American teaching zoology at the University of Tokyo, that they are "an ingenious way to teach *botany*." The subject of *botany* had been institutionalised in Japan as independent from the medical field for merely three years when Morse made this claim.[9] The clear distinction between the field of *materia medica* (or *honzōgaku*) and *botany* (or *shokubutsugaku*), which Morse ascribed to these objects, thus needs to be approached with a grain of salt.

The narrative arch for perceiving the application and appreciation of *botany* as part of a necessary seismic shift within the production of 'scientific knowledge' in nineteenth-century Japan and, more broadly, East Asia, typifies the methodological approach to visual materials of Meiji Japan.[10] In fact, I have relied on this method in my earlier work. According to this logic, these objects fill the role of evidence that bears witness to the 'successful' (read: accurate) transmission of a body of *botanical* knowledge from 'the West' to 'the East.' But this logic conversely highlights the absence of history from another side. That is, these objects also stood at another transection of intellectual and material histories untethered by the discourse of *botany* per se. Why these types of wood, and not others, for example? What stakes were involved in the production of these objects in Tokyo in 1878?

This chapter represents an attempt to redress the current narrative on these objects by taking seriously the motivations and challenges that existed at the place and time of the inception of these wood panels. I ask questions regarding the decisions made in the process of production, and the conditions of the Koishikawa Botanical Garden, to place these objects as evincing historical challenges of different

8 See my book, *The Premise of Fidelity: Science, Visuality, and Representing the Real in Nineteenth-Century Japan* (Stanford, CA: Stanford University Press, 2012).

9 Quoted in Nagata et al., "An Unusual Xylotheque with Plant Illustrations from Early Meiji Japan," p. 95. Also see, Edward S. Morse, *Japan Day by Day*, vol. 1 (New York: Houghton Mifflin Company, 1917), p. 396. Here, I take the 1875 decision to change the name of the *Koishikawa Botanical Garden* from *Igakukō yakuen* [Medicinal Herbal Garden of the Medical School] to Kyōikuhakubutsukan fuzoku Koishikawa shokubutsuen [Koishikawa Botanical Garden of Educational Museum] as a decisive shift. I will expand on this point more fully later in this chapter.

10 In this regard, Carla Nappi's work on Chinese medicine is particularly relevant. See Carla Nappi, *The Monkey and the Inkpot: Natural History and Its Transformations in Early Modern China* (Cambridge, MA: Harvard University Press, 2009). Also see her historiographic intervention in the article "The Global and Beyond: Adventures in the Local Historiographies of Science," in *ISIS*, 104 (2013): 102–110.

orders. What is the relationship between the two discourses, *materia medica* and *botany*, in the context of 1870s Japan? Was there a concerted effort on behalf of the government to replace one with the other? Or do elements of *materia medica* inform their practice of *botany* in epistemological and physically practical ways? And on a more historiographical level, to whom/what, why, and on what level, does the articulation of a power relationship matter between the two?

Specifically, I explore how these objects related to and functioned within the material culture of knowledge production. By shifting the perspective from the current location of the objects to one that privileges the origin of the objects and their conjunctural historical and political context, this chapter investigates how these objects embody economic functions and epistemic values that were paramount to the historical landscape of Meiji material culture. I thus situate the wood panels as objects located between two constitutive forces conditioning Koishikawa Botanical Garden and its employees. For the sake of clarity, I divide these forces into 1) economic and 2) epistemic, although the relationship between the two interlocks significantly.

Skills of Assemblage, Connecting Art and Economy

The Japanese term for fine art, *bijutsu*, was coined in 1872 in the context of a translation of the official invitation to participate in the Vienna World Exposition. The neologism *bijutsu* enabled the Meiji government to set apart selected objects from others, which immediately created a hierarchical order.[11] Many scholars have demonstrated how both international and domestic expositions played vital discursive roles as public sites for the evaluation and promotion of 'fine art.'[12] Conversely, these studies of historical conditions also reveal the fact that the image-makers in Tokyo did not – and could not – operate in the same way artists of the same period did in Paris, for example.

This historical incongruence poses methodological and historical challenges when we consider these wood panels. On what level can we characterise Katō's

11 See Suzuki Hiroyuki's *Kōkokatachi no jūkyūseiki: Bakumatsu Meiji ni okeru monono arukeoroji* (Tokyo: Yoshikawa Kōbunkan, 2003) and its English translation *Antiquarians of Nineteenth-Century Japan* (Los Angeles, CA: The Getty Institute, 2022). Kinoshita Naoyuki's publications explore objects excluded from the framework of 'fine art' within the visual culture of Meiji Japan. See especially, *Bijutsu toiu misemono* (Tokyo: Heibonsha, 1993); *Shashin garon* (Tokyo: Iwanami shoten, 1996); and *Yonotochūkara kakusareteirukoto: Kindai nihon no kioku* (Tokyo: Shōbunsha, 2002).

12 Satō Dōshin, *Modern Japanese Art and the Meiji State: The Politics of Beauty* (Los Angeles, CA: Getty Publications, 2003); Alice Tseng, *The Imperial Museums of Meiji Japan* (Seattle, WA: University of Washington Press, 2008); Norota Junichi, *Bakumatsu Meiji no biishiki to bijutsuseisaku* (Kyoto: Miyaobi shuppansha, 2015).

involvement in the production of these objects as belonging to the accepted category of 'artwork'? If we continue to describe Katō as an artist of these objects, what historical condition might be overlooked in such an approach, and what are the implications of such a decision? Similarly, Katō did not come to support the efforts at the Koishikawa Botanical Garden as a *botanical* illustrator. To what extent does his knowledge of *botany* matter in the production of these wood panels? I pose these not merely as semantic questions. In the context of this chapter, these series of questions highlight the scale of the challenges posed by methodological and discursive presuppositions that are easily overlooked.

Although there is no archive of Katō Chikusai as such, it is possible to compile a list of Katō's activities from historical records. In 1877, one year before the production of the Kew panels, Katō submitted a picture made with bamboo and *sawara* cypress to the First Domestic Expositions for the Promotion of Industry.[13] In a ranking list published in 1880, Katō is noted as an accomplished calligrapher, while his name appears as a notable painter in the Japanese tradition on the list of artists participating in the domestic exhibition of paintings in 1882.[14] Katō appears to have worked in at least three types of representations, eliciting distinct expectations from viewers: as an illustrator of plants for specialised purposes; as a painter in the Japanese tradition; and as a calligrapher. The inclusion of his names and varied skills in these lists become further complicated when we consider his own publication *Secrets of Paintings* (*Tansei hiroku*) in 1883. Here, Katō squarely presents himself as an experienced and trained painter in the Kanō pictorial tradition. These records present an image of Katō as resourceful in securing an income from his pictorial skills. The media he is associated with – calligraphy, Japanese painting, or, more specifically, Kanō school painting, and his illustrations at Koishikawa Botanical Garden – all use the brush to produce two-dimensional representations.

The question of disciplinarity on which I have expanded so far, can be summarised in the following observations: the transgressive and transdisciplinary

13 Submitted under Section 2, pictorial albums, Katō is named as the contributor for an item described as "coloured picture of flower and bird made with bamboo and *sawara* cypress." Tokyo kokuritsu bunkazai kenkyūjo, *Naikoku kangyō hakurankai bijutsuhin mokuroku* (Tokyo: Tokyo kokuritsu bunkazai kenkyūjo, 1996), 40. No description or price regarding the piece is given in the exhibition catalogue. Rather, a short description notes: "Studied skills of picturing under Kanō Tsunenobu, and in the end carried the pen name of *Kagawa*." *Meiji jūnen naikoku kangyō hakurankai shuppin kaisetsu*, included in Fujiwara Masato (ed.), *Meiji zenki sangyō hattatsushi shiryō* (Tokyo: Meiji bunken shiryō kankōkai, 1963), vol. 7, 73. Satō Dōshin asserts Katō submitted another item for the Second Domestic Expositions for the Promotion of Industry and received an award.

14 Miyata Uhei, *Tokyo shotaika raimeikyō* (Tokyo: Miyata Uhei, 1880), https://www.tobunken.go.jp/materials/banduke/806966.html; Otsuki Kinju (ed.), *Meiji jūgonen kaiga kyōshinkai shuppin gaka jinmei ichiran* (Osaka: Otsuki Kinju, 1883), https://www.tobunken.go.jp/materials/banduke/806931.html.

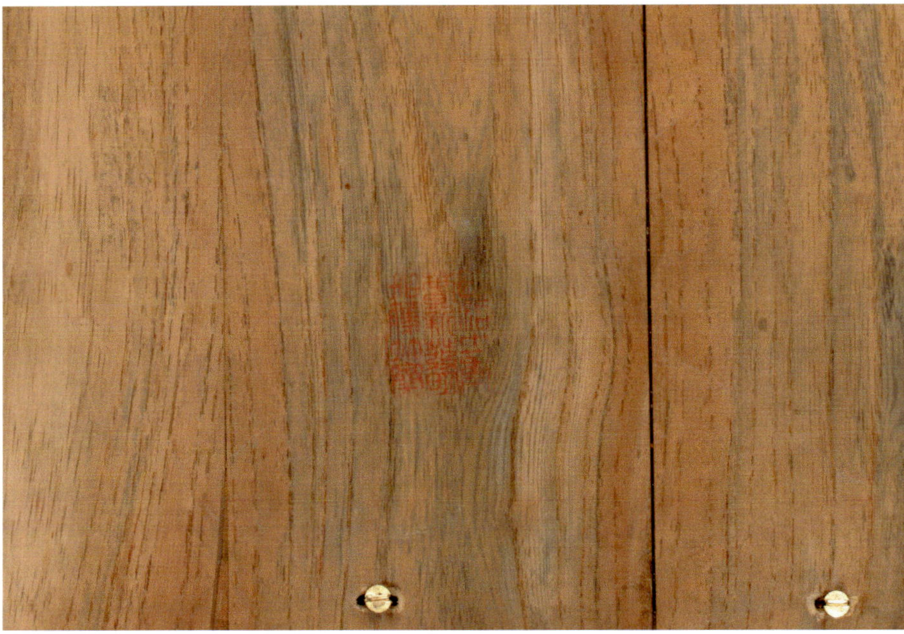

Figure 3.2: Recto of Figure 3.1. Red stamp noting "Meiji 11-nen, saishin hatsumei Katō Chikusai."

attributes revealed through examinations of these wood panels as a part of Meiji material culture, and the difficulty of situating Katō's work within a Eurocentric art historical taxonomy based on the putative concept of *medium*. The special red seal placed on the back of wood panels leads us to expand on the contextual and interpretive fields (Figure 3.2).

Word Matters: Patent for the Nation

It is in this initial step to probe the disciplinarity prescribed to this group of objects that the red seal placed on the back of each panel emerges as a significant sign. It reads "*Meiji 11-nen, saishin hatsumei Katō Chikusai,*" which can be translated roughly as "1878, the latest invention by Katō Chikusai." What would be the value and the point of producing a special stamp like this in 1878 and marking the object with it? What did Katō want to achieve with this stamp? What did he mean by the term *invention* (*hatsumei*) and why was it important to assert this?

Curiously, Takahashi Korekiyo (1854–1936), the first chief of the Japanese Patent Office (Tokkyokyoku), established in 1887, noted an intriguing insight in his autobiography: The 1885 process of drawing up a patent law was much influenced by the fact that the new Meiji government had had to withdraw earlier patent legislation in 1872.[15] Further, he noted that the failure of the earlier edict had resulted from two main causes: 1) establishing a panel of judges to review submitted applications would require hiring costly Westerners and appropriate translators; and 2) ordinary people found the concept of patent difficult to understand. Indeed, the extent of its failure was such that that the edict was withdrawn within a year of its promulgation. Between 1872 and 1885, Takahashi notes, the members of the rule-making body, the Councillor's Office (Sanjiin), persistently expressed strong opposition to the reintroduction of the patent law. At the same time, pressure on the Japanese government from 'foreign entities' to adopt some form of patent protection was increasingly palpable. The Kew panels were produced during the period between the first patent edict of 1871 and the final implementation of patent law in 1885. The word *hatsumei* included in the seal on the back of every panel thus serves as a keyword addressing the objects' economic potential on the one hand, and societal confusion the notion caused on the other.

Apparently, part of the difficulty in conveying the idea of patent lay in linguistic deficiency and perplexity. The word *hatsumei*, which I have translated as 'invention,' was interchangeable with *hakken* around this time. The term *hakken* would only solidify its meaning as 'discovery' in the contemporary English sense by the late nineteenth century. The first edition of James Curtis Hepburn's Japanese–English dictionary provides a clue to the historical and translational muddling. In the first edition, published in 1867, *hatsumei* is defined as "intelligent, ingenious, and clever." Within this entry, it is noted that its verb form means to "invent and discover" implicating the historical interchangeability of the two English verbs. The English word *discover* is defined by four Japanese verbs, "arawastz; mi-dasz; mi-tszkeru; hatszmei szru" while there is no entry for the Japanese term *hakken* altogether.[16] Furthermore, in the hugely popular Japanese translation of *Self-Help*

15 Takahashi Korekiyo, *Takahashi Korekiyo jiden* (Tokyo: Chikura shobō, 1936), pp. 218–219, http://dl.ndl.go.jp/info:ndljp/pid/1207485. The edict was issued on the seventh day of the fourth month of 1871 and withdrawn in the third month of 1872. To be sure, protection for copyright was issued as publication regulation (*shuppan jōrei*) in 1870, and the subsequent modifications of this law in 1872, 1875, and 1887 served as a way to protect rights, including photographic negatives.

16 The dictionary defines the English verb 'to discover' as "arawatz; mi-dasz, mitsukeru, hatszmei-szru" while the verb 'to invent' is defined as "hatzumei szru; takumu; takunamu." James Curtis Hepburn, *A Japanese and English Dictionary*, first edition (Shanghai: American Presbyterian Mission Press, 1867), pp. 25, 55, 96, and 164. At the same time, the English term 'monopoly' is defined as "kabu," and 'patent' is also defined as "kabu," a word conversely defined in the same dictionary as "[t]he stump of a tree. A government license, a guild, or fraternity of persons engaged in a same kind of business; a habit or custom to which one is addicted." For a recent exploration of how the term *hatsumei* functioned within regulations put forward by the Tokugawa *bakufu*, see Satomi Kobayashi, "Edojidai ni okeru hatsumei sōsaku to kenrihogo," in *Patent* vol. 61, no, 5 (2008), pp. 48–55. It is also noteworthy that it was James Curtis Hepburn, the compiler of this very English–Japanese dictionary, who first applied for copyright protection with the Meiji government. Hepburn's request in 1874 prompted the government to investigate more fully

(1859) by the Scottish educator and social reformer Samuel Smiles (1812–1904), published in 1870, the terms *hatsumei* and *hakken* remain undistinguished.[17] In this way, these wood panels begin to merge with broader context to inform and direct our understanding. The stamp Katō placed on the back of the wood panels allow us to see them as concrete objects to confer the fluctuating semantic field of *hatsumei* (invention) and its socio-economic ramifications.

One telling aspect of the Patent Office as established in Meiji Japan is that the office was administered by the Ministry of Agriculture and Commerce. In Europe and the United States, patent offices had historically operated under the Ministry of Justice as it concerned individual rights.[18] This idiosyncratic Meiji configuration illuminates the specific historical and political context: the newly formed national government saw the issue of patent in the light of national resources and agricultural and commercial activities.

Unsurprisingly, Fukuzawa Yukichi's seminal book *Conditions in the West* (Seiyō jijō, 1868) included a description of a patent office. In this popular book, based on his observations made in the UK, the Netherlands, France, and the United States, Fukuzawa, an avid proponent of Westernisation, refers to the patent office phenomenon as "*hatsumei kan*."[19] He described the institution as an important piece of infrastructure that evaluates the benefit (*eki*) for society at large (*yononaka*) as well as the individual (*hito*), in that order. The language and intent of the 1871 edict likewise carried the sense of national benefit prioritised over the protection of the rights and the income of the individual. Its first article, for instance, addresses industry rather than individual inventors.

both the copyright and patent issues. David Murray, an 'employed foreigner' at the Ministry of Education, mediated between the Meiji government and Hepburn. See Takahashi Korekiyo, "Wagakuni tokkyoseido no kiin," in *Kōgyō syoyūkenzasshi* vol. 32 (1908), republished online by the Institute of Intellectual Property under Takahashi Korekiyo ikōshū, https://www.iip.or.jp/chizaishi/korekiyo_ikosyu.html.

17 *Self-Help* and Fukuzawa Yukichi's *Seiyō jijō* (*Conditions in the West*, 1868) played significant roles in moulding the minds of the young Japanese, the former being used as a textbook. The translator of *Self-Help*, Nakamura Masao, uses *hatsumei* to translate the English noun 'discoveries.' For example, what is in Smile's original "although there are discoveries which are said to have been made…" is rendered as "*inishie yori gūzen ni sonokotowo hatsumei serito iukotoari…*" See Samuel Smiles, trans. Nakamura Masao, *Saigoku risshihen: genmei jijoron*, vol. 4 (Tokyo: Suharaya Mohei, 1870), p. 2., https://dl.ndl.go.jp/pid/1086561. In *Bunmei kaika naigai jijō* (Outline of Civilisations Domestic and Abroad), published in 1873, the author, Tōkō Gakujin, includes a section on the Patent Office and describes how possession of the patent for a craft can lead to a guaranteed prosperity for generations. See Tōkō Gakujin, *Bunmei kaika naigai jijō* (Tokyo: Higashinari Kamejirō, 1873), vol. *chū*, pp. 29–33, https://dl.ndl.go.jp/pid/761509.

18 Christopher Heath, "Intellectual Property and Anti-Trust," in Wilhelm Röhl (ed.), *History of Law in Japan Since 1868* (Leiden: Brill, 2005), p. 406.

19 Fukuzawa Yukichi, *Seiyo jijō,* vol 3. Appendix (Kyoto: Hayashi Yoshibei, 1868), p. 10, https://dl.ndl.go.jp/pid/993194.

> Article 1: The government will provide official patents for those new inventions (*hatsumei*) in the industries of chemistry, machinery, armoury, and textiles, as well as those outside of these industries that have not been recognised in the country. The government will also give a patent for those items which have improved on existing items and bring more convenience to society. These patents will have expiration dates.[20]

Fukuzawa's characterisation of a patent as a fundamental asset to build national wealth was cemented further with the 1887 establishment of the Japanese Patent Office under Takahashi Korekiyo.[21]

In addition, the concept of *hatsumei* emerged as a critical and ambiguous notion within the discussions held at the Council of Elders (Genrōin), a legislative assembly, in the context of patent legislation.[22] For instance, Mitsukuri Rinshō (1846–1897), a member of the Council and a specialist in French law, explained *hatsumei* first as being akin to the English noun 'discovery.' He noted that 'discovery' is an act of man finding new uses for natural elements, and listed electricity and steam power as examples. He then articulated a second meaning of the word *hatsumei* as a process of adding manpower to existing natural elements, and this, he noted, is the same as 'innovation' in English. He finally articulated a third type of *hatsumei*, which is one that improves upon a previous invention, for which the improver can only apply a patent for the proposed modification. Following Mitsukuri's articulation of the concept of *hatsumei*, the discussion moved to the difficulty in drawing the line between the original and the improvement, and the processes of evaluating such an improvement in textual or graphic formats. During this contentious and lengthy discussion, *hatsumei* was used as a term to encompass the three concepts of discovery, innovation, and improvement.[23] Katō's use of the term *hatsumei* in his

20 The edict consisted of twenty-eight articles. Dajōkan fukoku 175 jō, in *Hōrei zensho,* Meiji 4 (1871) (Tokyo: Naikaku kanpōkyoku, 1912), pp. 140–141, http://dl.ndl.go.jp/info:ndljp/pid/787951.

21 The first article of the Patent Law of 1885 asserts: "Those who have invented (*hatsumei*) useful things and want to retain a monopoly of the sales of the item should submit an application for a patent to the head of Ministry of Agriculture and Commerce. If the Ministry of Agriculture and Commerce deem the patent should be given, it will grant it to the applicant." The first patent was issued on 14 August 1885 to Hotta Zuishō for his invention of a lacquer-based, anti-rust paint to be used for painting the bottoms of metal ships. Curiously, Hotta was trained initially as a lacquer artist. This seeming 'multi-tasking' of an individual artist, quite typical of the period, demonstrates that 'art' and 'art-making' operated vastly differently in the context of early Meiji. Kaneko Kōichirō, *Senbaitokkyojōrei chūshaku* (Niigata: Miyajidō, 1885), p. 2.

22 *Genrōin kaigihikki*, no. 464 published in *Genrōin kaigihikki* (Tokyo: Genrōin, 1880), pp. 65–178, especially pp. 72–74, http://dl.ndl.go.jp/info:ndljp/pid/995129.

23 The difficulties related to the demarcation of these three concepts may be illustrated with the example of the spinning machine invented by Gaun Tokimune, known at that time as *garabō*. Gaun's machine had an external handwheel to twist threads from cotton tubes onto the reels. Gaun received the highest

stamp of 1878 sheds light on the on-going and unresolved role of the patent in the emergent modernisation, and the uncertainty about what such legislation should protect, and in what manner.

The Usefulness of Timber

Let us extend the analysis of economic value articulated on and via the wood panels further. Between the failed attempt to introduce the patent law in 1871 and its implementation in 1885, the Meiji government was actively promoting another seminal project to buttress their nation-building efforts: international and domestic expositions.[24] It orchestrated a total of five Domestic Expositions for the Promotion of Industry, the first of which took place in 1877. The fact that during the first domestic exposition, over thirty per cent of the submitted items received some form of award from the organising committee attests to the eagerness of the officials to use the platform for the purpose of identifying potential economic opportunities.[25] The government was even more active in participating in exhibitions abroad. According to one account, governmental and private entities from Japan officially participated in at least nineteen international expositions between 1872 and 1885.[26] In fact, from the record kept at Koishikawa Botanical Garden, we learn that some of the Kew panels were exhibited at the 1884 Exposition Internationale d'Horticulture in St. Petersburg, Russia.[27] The fact that these objects served as carriers of values within the international exposition circuit brings us to

honour for his invention at the First Domestic Expositions for the Promotion of Industry. Because he did not have patent protection, however, many entrepreneurs copied his machine, and Gaun lived in poverty. The 1877 machine was described as *hakken* (a discovery), because it did not exist before his submission at the exposition, while a modified version, which used a pedal instead of a handwheel, is described as an "improvement." See *Komonroku,* vol. 96 (April 1885), Ministry of Agriculture and Commerce, n.p., http://www.archives.go.jp/exhibition/digital/hatsumei/contents/photo.html?m=11&ps=1&pt=5&pm=1.

24 The classic text on the history of museums and exposition in Japan is Yoshimi Shunya, *Hakurankai no seijigaku: manazashi no kindai* (Tokyo: Chūōkōron, 1992). See also Seki Hideo, *Hakubutsukan no tanjō: Machida Hisanari to Tokyo Teishitsu hakubutsukan* (Tokyo: Iwanami shoten, 2005). Shiina Noritaka offers an intriguing study of historical episodes related to the establishment of museums and expositions in *Meijihakubutsukan kotohajime* (Tokyo: Shibunkaku shuppansha, 1989). For a detailed account on the processes of constructing an image of 'Japan,' and the production of writing its histories at international and domestic expositions, see Itō Mamiko, *Meiji nihon to bankoku hakurankai* (Tokyo: Yoshikawa kōbunkan, 2008).

25 The contrasting gap between the governmental officials' and the populace's expectations for the exposition is described wittily in the sculptor Takamura Kōun's recollections of the early Meiji years. Takamura Kōun, *Bakumatsu ishin kaikodan* (Tokyo: Iwanami shoten, 1995), pp. 122–128.

26 Satō, *Modern Japanese Art,* 108.

27 Murata Hiroko, et. al., "'Meiji jūshichinen jūnigatsu Tokyo...'," pp. 213–214.

examine what types of value, according to the Meiji government, were associated with these woods. Careful consideration of the interlinked activities among the government's participation in expositions, *Useful Timbers: A Quick Reference* (1874), a governmental publication on the projected economical and material values of timber, and these wood panels lead us to the monetary and commercial profit imbued into the Kew panels.

In fact, after their successful participation at the 1873 Vienna Exposition, the government's Exhibition Bureau (Hakurankai jimukyoku) put forward a series of books through which they articulated the parameters of usefulness as an operative term that played a key role in governmental projects. At the same time, the structural configuration and thus significance of the Exhibition Bureau shifted after the Vienna Exposition, when its office was moved from the remit of the Central Council of the Ground Council of State (Dajōkan sei'in) to that of the Ministry of Education (Monbushō).

If ordinary people and the new Meiji politicians were initially perplexed by the question of how the abstract notions of patent fit into their changing society, then we might imagine that the uses of wood and its applicability as a material would not have caused a similar level of confusion. A large variety of woods had been utilised in everyday contexts for centuries, and the associated businesses of forestry and crafts had long been established and were familiar to most people. The Meiji government's push to highlight the usefulness of wood, then, was not particularly novel or inventive per se. In fact, in the last decades of the Tokugawa period, the shogun and domanial lords made concerted efforts to encourage greater attention to and appreciation for their local products to promote their economy. What distinguished the Meiji government's effort was its foregrounding of certain types of wood as 'useful' building materials for larger commercial projects, especially for export purposes.

Works by Tanaka Yoshio (1838–1916) serve as solid resources when tracing the manner of how 'useful' aspects of materials became highlighted. Tanaka was one of the most active members of the Exhibition Bureau. Before working there, he had been trained as a student of *materia medica* by Itō Keisuke, whom he met in 1856 at the age of eighteen. In 1863, Itō sent Tanaka to Edo (renamed Tokyo in 1868), to study at the Academy for Western Studies (Yōsho shirabejo). Tanaka had been a member of the Japanese delegation taking part in the Paris Exposition in 1867, selected for his familiarity with the discursive content and the challenges of *materia medica*. Another aspect of Tanaka that appealed to the selection committee was his activities regarding exhibition and collection practices in the context of the Society of One Hundred Tasters (Shōhyaku-sha), the group of *materia medica* scholars Itō led in Owari domain.[28] He then joined the new Meiji government, most notably as a member of the Exhibition Bureau. Tanaka played a leading role in the government's participation in the 1873 Vienna Exposition, the 1876 Philadelphia Exposition, and the first Domestic Expositions for the Promotion of Industry in 1877.[29]

In November 1874, the Exhibition Bureau published *Useful Timbers: Quick Reference* (Yūyō mokuzai shōran).[30] At that time, the Bureau would have been preparing for the Philadelphia Exposition as well as the first domestic exposition, but this booklet was meant for the Japanese public. *Useful Timbers* describes one hundred trees available domestically. What characterises this book is the fact that pieces of physical elements of the timber are pasted onto each page: two slices – flat-sawn and quartersawn – of the trunk (Figure 3.3). Four years prior to the production of the wood panels now in Kew's Economic Botany Collection, *Useful Timbers* ventured to provide samples as part of a reference book. The use of physical wood in this publication is striking, especially considering the production of the wood panels.

In terms of format, *Useful Timbers* is similar to commercial sample books produced, for instance, by the textile industry.[31] It departs significantly, however, from contemporary publications on trees or plants, which relied heavily on pictorial representations in the woodblock format.[32] Especially in cases where the hue of the trunk plays a key role in distinguishing the types (for instance, in the first two entries: "*sugi* whose heartwood is red" and "*sugi* whose sapwood is white") the effectiveness of using the physical pieces, rather than pictorial representations, takes advantage of its format. The introduction to *Useful Timbers* notes the book's intentions:

> The trunk is the most useful part of the tree, from building houses and ships with large pieces to producing smaller items of utility such as desks, chairs,

28 A catalogue published for a commemorative exhibition organised in honour of Tanaka's career offers more information: Dainihon sanrinkai (ed.),*Tanaka Yoshiokun nanaroku tenrankai kinenshi* (Tokyo: Dainihon sanrinkai, 1926). For the activities of Shōhyaku-sha, see Fukuoka, *The Premise of Fidelity.*

29 Kinoshita Naoyuki, "Daigaku nankō bussankai nitsuite," in *Gakumon no arukeorojī*, http://umdb.um.u-tokyo.ac.jp/DPastExh/Publish_db/1997Archaeology/01/10700.html. Suzuki, *Kōkokatachi*, 121–123.

30 Hakurankai jimukyoku, *Yūyō mokuzai shōran* (Tokyo: Hakurankai jimukyoku, 1874). Originally founded in the second month of 1872 within the Central Council (*Sei'in*) of the Ground Council of State (*dajōkan*), the Exhibition Bureau was critical to preparing for the Vienna World Exposition. The Museum Bureau (*hakubutsukyoku*), on the other hand, was initially founded in 1871 within the Ministry of Education (*monbushō*). The Exhibition Bureau absorbed the Museum Bureau in 1873. In 1875, the Exhibition Bureau was transferred from the Ministry of Education to Home Ministry (*naimushō*), which then moved it to the Ministry of Agriculture and Commerce (*nōshōmushō*) in 1881. Suzuki, *Kōkokatachi*, Chart 1, pp. 13–14.

31 See for example, *Shoorimono shimahon shūchō*, 1872, https://dl.ndl.go.jp/pid/2541098.

32 See for example, Abe Rekisai, *Sōmoku sodategusa,* published in 1876, https://dl.ndl.go.jp/pid/1911130.

> boxes, combs, chopsticks, and abacuses. We now display a few types of trunks in our museum to show them to the public. However, it is difficult to always carry these wood samples as a tool to consider their names and utilities, even when the sample pieces are small. Therefore, we have made thin slices of many types of trunks cut horizontally and vertically and pasted them into a book format. This small booklet fits inside one's hand and includes samples of one hundred trunks. At a glance, one can know their names, the famous regions that produce these woods, and their appropriate uses.[33]

The idealised image of the general public evoked by this introduction is intriguing: it is assumed that readers would want to know the commercial uses of the trees they see as they go around their neighbourhood and villages. Although this may not be a realistic expectation, the publication clearly demonstrates the government's desire to imbue the 'useful' aspect of wood as general knowledge. The introduction also asserts didactically that, in the case of wood, the process of verifying its type was made more challenging by its sheer physicality and immobility.

The three different uses identified in *Useful Timbers* are: 1) construction of buildings; 2) construction of ships; and 3) manufacture of smaller-scale commercial products such as bowls. The significance of the 'usefulness' in the title is thus articulated: these one hundred types of trees are 'useful' to the extent that they provide the raw material for these purposes and products. The medicinal properties of these specimens do not have a place in this book. Instead, *Useful Timbers* stresses the value of wood as important industrial material, as *timber* in short, and underscores commercial use including international trade.[34]

The order in which the tree species appear in the book seems to have been determined by the prevalence of the tree and its broader variant common names. For instance, the book begins with what is designated as "*sugi* [Japanese cedar] whose heartwood is red." It is then followed by four variants of the type: "*sugi* whose sapwood is white," "*Yaku-sugi*," "*Kurobe-sugi*," and "*Jindai-sugi*." This grouping based on a broader common name allows for easy reference among the variants. The

33 Hakurankai jimukyoku, *Yūyō mokuzai shoran* (Tokyo: Hakurankai jimukyoku, 1872), https://dl.ndl.go.jp/pid/842435, n.p. The introduction further notes that the pasted pieces were taken from a tree grown in the region whose name appears first on the list of names provided on each page.

34 During the Tokugawa period, especially the nineteenth century, some of the objects included in canonical works such as *Bencao Gangmu* came to hold a significant commercial value, especially in trade relations between domains. Metals received increased attention in this regard, as the quality of minerals and metals affected the quality of farming tools. It is also essential here to underscore that the same criterion of usefulness continued to be applied to timber as late as 1889, on the occasion of the Exposition Universelle. See Maeda Kenkichi, *Nihon yūyōmokuzai hyōhon kaisetsu* (Tokyo: self-published, 1889).

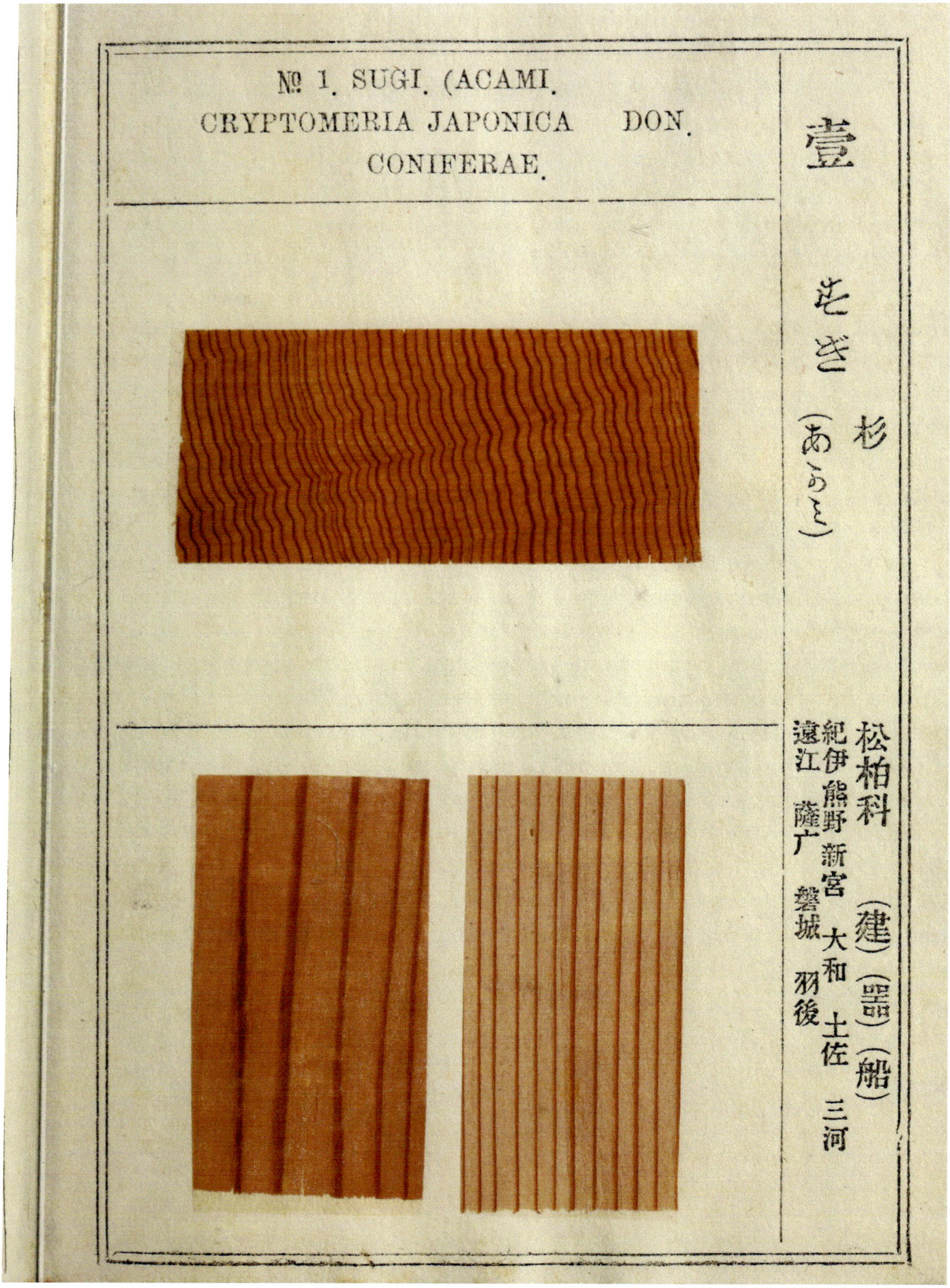

Figure 3.3: A page of Sugi (Acami) from Exhibition Bureau (Hakurankai jimukyoku), *Useful Timbers: Quick Reference* (*Yūyō mokuzai shoran)*, published in 1874, woodblock print, and glued wood segments.

access to actual texture and smell, made immediate by this format, stands in stark contrast to the transfer of knowledge based on texts and two-dimensional images. On the page for "*sugi* whose heartwood is red," the description makes clear that this type of *sugi* is useful for all three purposes. In contrast, no uses are assigned to "*sugi* whose sapwood is white." It is instead accompanied by a note, explaining that "its solidity and density are inferior to *sugi* whose heartwood is red."[35]

The trees represented in this book are treated equally in the sense that each has one page dedicated to it. Despite the simple layout, each page consists of seven components: the common Japanese name of the species in questions; its Latin binominal name; designated uses of the wood selected from the three options noted above; the broader family name of the tree in Japanese; the physical characteristics of the tree; the names of regions well-known for this particular type of tree; and the species' regional name(s).[36] By way of example, let us look at the first page "*sugi* whose heartwood is red" (Figure 3.3). On the right side of the page, the text notes, in horizontal order, a Chinese character for the numerical one, followed by the word *sugi* in cursive *hiragana* script. Under this, it inserts *sugi* in Chinese characters on the right, then the word *akami*, meaning red heartwood in cursive *hiragana* in parenthesis. At the top of the page, in horizontal order, the text is rendered in the Roman alphabet: a numerical one, then capital letters spelling SUGI. In parenthesis, it notes in capital letters ACAMI, which appears to be the romanisation of the Japanese name *akami*, as noted in the right-hand column. This is then followed by the tree's binominal name, Cryptomeria Japonica Don., then the class to which *sugi* belongs, *Coniferae*, within plant taxonomy. Returning to the right-hand column of the page, the Chinese characters grouped towards the bottom half of the page begin with *Shōhakuka*, a Japanese term that refers to evergreen trees. Under this, there are three terms, each printed within parenthesis – (architecture), (bowls), and (ships) – in horizontal order. The criteria used here are identical to the uses identified in the introduction, and the order in which these are noted seems to reflect the order in which this type of tree is most useful.

The pairing of Latin names with (a) common Japanese name(s), seems out of place at a time when very few members of the public were familiar with the Roman alphabet.[37] It does, however, reflect the somewhat overambitious intention behind this publication: that ordinary people might take up botanical explorations. In fact, the introductory note recommends another publication, *An Introduction to Botany* (Shokugaku senkai), for further study. We will return to this book in the next section.

35 Hakurankai jimukyoku, *Yūyō mokuzai*, n.p.

36 Hakurankai jimukyoku, *Yūyō mokuzai*, n.p.

37 Although this can also be understood as a rhetorical device that lends authority precisely because it deploys language remote from the general understanding of its intended readers.

A consideration of two items in the records of exhibited items in the timbers section at the First Domestic Exposition for Promotion of Industry (1877) further contextualises *Useful Timbers*. On the occasion of this first domestic exposition, Tanaka asked Itō Keisuke to serve as the leading judge for this section.[38] Needless to say, this event was one of the main projects for the Exhibition Bureau when Tanaka began to work there in 1871. Among the numerous types of *sugi* timbers and seedlings exhibited, the records mention a publication titled *Collection of Useful Plants and Their Samples* (Yūyō shokubutsu ruishū oyobi mihon) as well as one called *Framed Timber of Various Types* (Mokuzairui gaku).[39] The latter, the description informs us, can be none other than the publication we just examined: *Useful Timbers*. The description of the first publications sheds an interesting light on our inquiry:

> Plants that grow between heaven and earth offer innumerable uses. There is not a single object that does not contain some plant elements, from clothing, food, and the building of houses, ships, and wheels, to various types of utensils and items for storage. However, if people were asked to evaluate the qualities of wood, they are unable to understand its characteristics and develop its unknown potentialities. Therefore, we collected outstanding examples of plants that are widely used today. We called them *Collection of Useful Plants*, placed them into categories, and added their Japanese, Chinese, and Western names. We are now also selling samples of these useful plants so that everyone can collect these small pieces and have them to hand.[40]

The rationale for making these samples available for purchase clearly overlaps with the intentions of *Useful Timbers*. The Exposition Bureau's driving idea must have been that educating people about the uses and names of various trees was the first order of business. Names, naming, classifications, and an awareness of the variant appearances of trees were seen as part of this learning process. Publications such as *Useful Timbers*, as well as the samples that were sold at the Exposition, were solutions to overcome impediments to the Bureau's enlightening mission. Although no definitive record has so far been found, it makes historical and material sense that wood panels such as the ones now kept at Kew had their origin in *Collection of Useful Plants and Their Samples*.

38 To be exact, the section of timbers belonged to the larger department of Agriculture. Itō Keisuke also served as a chief judge for the Gardening Department, see Sugimoto Isao, *Itō Keisuke* (Tokyo: Yoshikawa kōbunkan, 1960), p. 250.

39 In *Meiji jūnen naikoku kangyōhakurankai shuppinkaisetsu jumoku*, reprinted in Fujiwara *Meiji zenki*, p. 51.

40 Fujiwara, *Meiji zenki*, p. 51.

Indeed, from the diaries of Itō Keisuke, we learn that Itō asked Katō to produce at least 160 wood panels. Of them, at least fifty panels were purchased by Katō Hiroyuki, the president of the Department of Science, Law, and Letters at the University of Tokyo, for fifty yen, which equates to one yen per panel.[41] To give a sense of the economic value attached to the panel, the cost of a portrait photograph using the collodion process, which was becoming a popular commodity, was about one seventh of one yen or fifteen *sen*. This cost amounts to the average daily wage of a male farmer. In other words, each of the wood panels was expensive, amounting to seven working days of farming labour.[42] The sale of 160 of them, therefore, accrued a considerable amount of income.

The relationship between the publication *Useful Timbers,* the material form provided by the wood panels, and the exposition activities further underscore the political significance the government placed on orchestrating expositions and exhibitions. Here, the relationship between the thoughts of Fukuzawa Yukichi, who did not hold a political appointment in the government, and the way governmental officials reappropriated Fukuzawa's articulation of usefulness in their practice warrants attention. Most significantly and unsurprisingly, Fukuzawa Yukichi deployed the concept of 'usefulness' in explicating the role of expositions in his *Conditions in the West* (Seiyo jijō). In volume one, he focuses on the exhibition infrastructure found in 'the West,' including various types of museums as well as world expositions. In describing the motivations for hosting expositions, he notes:

> As described in the previous entry on "museums," each country establishes museums that collect articles, old and new, from around the world. However, the arts and crafts of each nation (*kuni*) develop daily, and new inventions (*hatsumei*) follow. There is always something new. For this reason, it often happens that what was considered a rare and precious item in the past is seen as stale today, and that yesterday's clever item becomes useless today. Therefore, in metropolitan cities of the West, they organise an assembly of products from each nation to gather notable products, useful instruments, old and rare items through a notification, and show these to people from around the world. These are called expositions.[43]

41 Itō's request was made between 2 and 22 June 1878. Murata et. al, "'Meiji jūshichinen jūnigatsu…'," pp. 214–216.

42 Based on Itō's diary entries, some scholars assert that the production of the wood panels was overseen by Katō who managed several artisans, as it would be challenging to produce such a number by himself. See Nagata et. al, "An Unusual Xylotheque," p. 95.

43 Fukuzawa Yukichi, *Seiyo jijō* (Tokyo: Hayashi Yoshibe, 1868), http://dl.ndl.go.jp/info:ndljp/pid/993191, pp. 42–43. Kume Kunitake, a member of another delegation, sent to the US and Europe in 1871, made similar observations when describing the Vienna Exposition. He imagines "benefit" (*rieki*) as something

One of the urgent motivations for hosting an exposition, according to this text, is to publicly update and confirm the status of exhibited items as either useful or useless. Fukuzawa's repeated use of the word 'nation' re-inscribes his view of the nation as the primary beneficiary of 'useful items.' This useful-useless dichotomy applied to articles is precisely what the government deployed in dealing with the broader material and visual culture of early Meiji Japan.[44] Meiji politicians, in fact, steadily amplified Fukuzawa's perspective on the purpose of expositions. Major governmental bureaucrats such as Okubo Toshimichi, Sano Tsunetami, and Machida Hisanari, adapted the educational power of public showings, such as exhibitions and expositions, as a suitable platform for the transfer of knowledge and values. Similarly, they did not question the view that international expositions and museums served the 'civilisational' role of strengthening national wealth and its competitive position in broader international trade networks.

As mentioned earlier, timber had played a central role as a construction material for centuries, and thus its usefulness was indisputable. What was novel in the Meiji period was the governmental effort to articulate to whom and how usefulness was defined. Namely, the government made a concerted effort to assert that particular types of timber were useful for the export business and the government to achieve their political vision for the new nation. In this light, the dissemination of knowledge concerning the commercial potential of timber became a matter of urgency. That all but two of twenty-six panels at Kew are included in *Useful Timbers*, therefore, carries a particular economic significance for the government.[45] That is, when Koishikawa Botanical Garden produced the wood panels in 1878, the Exhibition Bureau had already identified these selected types of wood as raw materials that could bring trade profits. It is telling that in the report of the First Domestic Exposition for the Promotion of Industry, Gottfried Wagner, an advisor to the government committee,

that should trickle down from nation to individuals, with an emphasis on material wealth. Quoted in Yoshimi, *Hakurankai*, 118.

44 This proved particularly decisive in reconfigurations of antiquarianism and the establishment of art history. For a detailed study on the activities of antiquarians around this period, see Suzuki, *Kōkokatachi*. Kinoshita Naoyuki takes the treatment of stone axes in various survey publications on 'art history' in Japan to demonstrate compellingly that the concept of 'use' played a significant role in excluding or including objects from the discourse. In particular, he points to a speech by Imaizumi Yūsaku at the Dragon Pond Association in 1884, in which Imaizumi deployed the concept of 'use' as the definitive criterion to distinguish between fine art and antiques. See Kinoshita Naoyuki, "Nihon bijutsu no hajimari," in *Yonotochūkara kakusareteirukoto*, p. 330. Also see Satō, *Modern Japanese Art*, especially Chapter 3 "Art and Economics." Takagi Hiroshi offers analyses on the issue of periodisation of 'Japanese Art History' in "Nihon bijutushi no seiritsu shiron," in *Nihonshi kenkyū*, vol. 400 (1995): 74–98.

45 The two exceptions are: *nikkei* (cinnamon), and *saikachi* (honey locust, *Gleditsia japonica*).

recommended refocusing on the timber industry, rather than on the export of timber, because of exorbitant shipping costs.[46]

Furthermore, Koishikawa Botanical Garden also incorporated the category of "useful timbers (*yūyō mokuzai*)" for the basic cataloguing of items at the site. The category appeared for the first time in 1875, three years prior to the production of the wood panels, and lists twenty-four items. The category then is deployed every year in their practice until 1882, when the term "useful timbers" was replaced with "*timber samples* (*mokuzai hyōhon*)."[47] In other words, the urgency of asserting the idea of usefulness was not only targeting the general population. The Botanical Garden itself relied on the concept and the value associated with it to concentrate on timber as a resource worthy of *their* attention. In view of the seven-fold increase in the numbers recorded under the category of "useful timbers" – from twenty-four in 1875 to 170 in 1879, the institutional and governmental attention and labour that went into the articulation of "useful timbers" must have been extraordinarily intense.

It would, moreover, be misleading and premature to characterise the motivations for producing the wood panels as exclusively economic and commercial. As noted in the introduction, there were two major constitutive forces at Koishikawa Botanical Garden and associates such as Itō Keisuke and Tanaka Yoshio. Let us turn to the publications on the *nomenclature* of trees by Itō and Tanaka to analyse the localised yet forceful epistemic negotiations taking place within the specialist study of trees.

Epistemological Stakes on the Trees

Today, *botanical garden* (*shokubutsuen*) is a familiar term. But at the time of the Meiji Restoration, no sites called *shokubutsuen* existed yet, and the process by which Koishikawa Botanical Garden became known as such further illuminates the circumstances of the production of the Kew panels and the values prescribed to them.

Established in 1711 as a medicinal garden for the Tokugawa shogunate, its primary function was to cultivate plants to be used in the preparation of medicine in the manner taught in *materia medica*. Following the 1868 restoration, the garden went through various hands. In 1877, the garden became part of the newly established

46 Gottfried Wagner, Asami Tadamasa trans., *Meiji jūnen naikoku kangyō hakurankai hōkokusho* (Tokyo: Naikoku kangyō hakurankai, 1877), p. 64, https://dl.ndl.go.jp/pid/801863. Wagner served as an advisor to the Administrative Office for the 1873 Vienna Exposition, and the 1876 Philadelphia Exposition. "Dokutoru Wagunerushi den," in Tanaka Yoshio, Hirayama Narinobu, *Ōkokuhakurankai sandō kiyō* (Tokyo: Moriyama Shunyō, 1897), pp. 53–72, https://dl.ndl.go.jp/pid/801730. "Waguneru-den," in Dainihon yōgyōkyōkai, *Nihon kinsei yōgyōshi* (Tokyo: Dainihon yōgyōkyōkai, 1922), pp. 1500–1502, https://dl.ndl.go.jp/pid/970706.

47 Murata et. al, "'Meiji jūshichinen jūnigatsu…'," pp. 240–246.

University of Tokyo.[48] Before that, between 1868 and 1877, the garden was known as *Koishikawa Yakuen*, roughly translated as Medicinal-Herb Garden in Koishikawa.[49] It was placed under the administrative responsibility of the Bureau of Local Products (Bussankyoku), an office under the Ministry of Education.

When the garden was reassigned to the Department of Science (Rigakubu) of the University of Tokyo in 1877, disagreement within the university persisted over whether the medical school would be its appropriate affiliation.[50] This debate was largely the result of the prevailing historical association of plants with *materia medica,* and the centrality of herbal remedies in medical practice. The wood panels, thus, are embedded within these shifts in overarching goals involving the government as well as the structure of a 'university,' government's central institution to disseminate and refine knowledge. Knowledge of trees at Koishikawa Garden was purposed for two distinct goals: the first fundamentally as an extension of medicinal uses, and the second as a significant resource for nation-building. At first sight, Itō's position at the Garden appears to be uncertain. Morse's comment in 1878 that these wood panels were wonderful teaching tools of *botany* thus need to be situated within this discursive and epistemic instability.

Itō began his official affiliation with Koishikawa Botanical Garden in 1875.[51] At that time, he was seventy-two years old.[52] His first publication during his tenure was in 1877, the year the garden became part of the University of Tokyo. Titled in

48 The University was founded in 1877 by merging two pre-existing schools: Tokyo Medical School (Tokyo igakkō) and East School of Daigaku (Daigaku higashikō), and Tokyo Kaisei School (Tokyo *Kaisei* gakkō). In 1885, the university absorbed the Tokyo Law School (Tokyo hōgakkō), which was administered by the Ministry of Justice, and the Technical Art College (Kōbu daigakkō), which was administered by the Ministry of Technology, to re-establish itself as a general university known as the Imperial University (Teikoku daigaku) in 1886.

49 During the Tokugawa period, the same garden was known as the Koishikawa Goyakuen (roughly translated as the Shogunal medicinal-herb garden in Koishikawa). In the early years of the Meiji period, the honorary prefix '*go*,' which indicates the affiliation with the shogunal family, was dropped. In Itō Keisuke's brief history of the garden, he notes that, in 1868, the administrative body of the garden shifted from the shogunal to the city of Tokyo, specifically its hospital under the name *goyaku'en*. In 1869, the garden was called the medicinal-herb garden of the Medical School (Igakukō yakuen) when it was absorbed by the Tokyo Medical School. The Ministry of Education then absorbed the garden in the seventh month of 1871, and, three months later, it was jointly administered by the Exposition Bureau under the name Koishikawa shokubutsu'en. Itō Keisuke, "additional notes (*fuki*)," in *Catalogue of Plants in Koishikawa Botanical Garden* (Tokyo: University of Tokyo, 1879), vol. 2, p. 1.

50 Oba Hideaki, "Nihon no honzōgaku no ayumi to koishikawayakuen no rekishi," in Oba Hideaki ed., *Nihon shokubutsukenkyū no rekishi: Koishikawashokubutsuen sanbyakunen no ayumi* (Tokyo: University of Tokyo Museum, 1996), pp. 21–49.

51 For details regarding Itō's appointment, see Sugimoto, *Itō Keisuke*, pp. 225–233. The changes in his affiliations demonstrate the complexity of untangling and regrouping pre-existing disciplinary and institutional practices that took place in the early Meiji period on a governmental level.

52 By the time Itō joined the garden, moreover, the subject of his expertise, *materia medica,* had become the official responsibility of the Bureau of Local Products (Bussankyoku). In other words, the Meiji

English as *Catalogue of Plants in Koishikawa Botanical Garden*, it listed the plants grown at the garden in Chinese, locally known Japanese, and Latin binominal names. This two-volume publication includes only textual descriptions. Four years later, in 1881, Itō published the three-volume *Figures and Descriptions of Plants in Koishikawa Botanical Garden* with illustrations by Katō Chikusai. It is an expanded, illustrated version of the 1877 catalogue with additional regional names, and descriptions that include histories of how the plant came to the garden, observations on its growth, and medicinal uses. The production date of the Kew wood panels thus falls between these two publications. Through cross-referencing, we learn that all twenty-six kinds of wood represented in the panels were growing in the Koishikawa at the time of their production.

In these attempts to disseminate knowledge of plants, Itō maintained an unwavering methodological approach: he began with plants he had direct physical access to. He then identified the plant by providing its name(s) in local variants, its name(s) as listed in canonical books on *materia medica* in Chinese and Japanese, and its Latin binominal name(s), often directly asserting the impossibility of knowing all names and their variants. In the introductions to both the 1877 and the 1881 catalogue, Itō lamented the confusion caused by names and naming, claiming that one of the intentions of these publications was to "make it as clear as day (*ichimoku ryōzen*)."[53] In the *Catalogue of Plants in Koishikawa Botanical Garden*, he named his sources for the Latin names, referring to the Swiss scholar Augustin Pyramus de Condolles, the Dutch botanist Cornelis Antonie Jan Abraham Oudemans, the German naturalist Philip Franz von Siebold, and the French botanist Paul Amédée Ludovic Savatier, emphasising the divided state of botanical nomenclature. In *Figures and Descriptions of Plants in Koishikawa Botanical Garden*, Itō noted "although I consulted scholars such as German Philip von Siebold and French Savatier, often we invent new names, or hear new theories that necessitate the modification of names. Therefore, we must keep editing these names."[54] Curiously, in the introduction to the 1881 catalogue, he also asserted, self-reflectively, the impossibility of reaching anything approaching perfection in accumulating botanical knowledge, given the uneven and fragmented condition under which knowledge about plants

government had recognised the economic potential of *materia medica*, which, until then, had been primarily a subject for medical professionals.

53 Tokio Daigaku, *Catalogue of Plants in Koishikawa Botanical Garden* (Tokyo: Scientific Department of Tokio Daigaku, 1877), vol. 1, n.p., https://dl.ndl.go.jp/pid/832378; and University of Tokyo ed., *Tokyo Daigaku Koishikawa Shokubutsuen sōmoku zusetsu* (Tokyo: Maruzen, 1886), vol. 1, Kato's name appears on the bottom left corner outside the frame of illustration, which is magnolia on page 21, https://dl.ndl.go.jp/info:ndljp/pid/832374.

54 Itō Keisuke, "hanrei" in University of Tokyo ed., *Tokyo Daigaku Koishikawa Shokubutsuen*, vol. 1, n.p., https://dl.ndl.go.jp/pid/832374.

becomes accessible.[55] A historical description such as this gives an insight into the ongoing and concomitant roles *materia medica* and *botany*, or *honzōgaku* and *shokubutsugaku*, played in the minds of Itō and, more broadly, within the process of knowledge production at Koishikawa Garden.

The increasing scale and scope of the above-mentioned publications – from a simple list of names in 1877 to a detailed and elaborate compilation in 1881 – is typical for Itō's long career. As he himself developed from a physician trained in Chinese medicine to a scholar of *materia medica*, and finally to a specially appointed Professor (*Rigakubu ingaikyōju*) at the University of Tokyo in 1881, his approach to the accumulation of knowledge remained steadfast and resolute.[56] Itō actively took part in the tidal transformation around and for the study of plants – modifying the discursive framework from *materia medica* that originated in China, to *botany* built upon the European taxonomic systems – through his recursive study of the systems of nomenclature. While this epistemic shift led to further inquiries and questioning among scholars of *materia medica*, it also necessitated coming to terms with taxonomic disputes among *botanical* scholars. The epistemic debate documented in the publications included in this chapter, therefore, not only took place between *materia medica* and *botany*, but also considered disagreements within the field of botanical science as discussed below.

Indeed, the structured method attested in Itō's *Figures and Descriptions of Plants in Koishikawa Botanical Garden* parallels his work back in Nagoya when he led the Society of One Hundred Tasters for four decades.[57] While the government increasingly came to see timber as a resource for international trade, Itō's own approach to the study of plants did not change radically to conform to this shift in official attitude. In other words, his systematic exploration of plants did not become irrelevant or, more to the point, "useless" by the broader change in the

55 According to Itō, this is the case even if one continuously revises and modifies their knowledge. In this context, he mentions the numerous regional names, the plants that did not survive even when acquired by the Koishikawa Botanical Garden because of their temperament, and the vast number of plants that grow in inaccessible areas. University of Tokyo ed., *Tokyo Daigaku Koishikawa Shokubutsuen*, n.p.

56 Because Itō was already working for the Garden in 1875, his new appointment at the University of Tokyo when the Garden was absorbed to the University enabled him to oversee the operations and research at the Garden. Although the title of professor implies a teaching component, it is important to underscore here that Itō did not teach botany at the university. At this time, the subject of *botany* was taught at higher education level at the Hitotsubashi University, taught by Yatabe Ryōkichi. Sugimoto, *Itō Keisuke*, pp. 243–244.

57 Itō began his pursuit of *materia medica* in Owari by following up on his teacher Mizutani Hōbun's work *Clarification on the Names of Things* (1809). Itō's *Nominal Differentiations in Western Materia Medica* published in 1829 expanded Mizutani's work with the application of the Linnaean binomial system. Itō then began to compile albums of local plants, which included ink rubbings, cut-outs of woodblock printed illustrations from other publications, as well as his own sketches. See my book, *The Premise of Fidelity*, especially Chapter 2, "Ways of Conceptualizing the Real: Scripts, Names and *Materia Medica*."

direction of governmental interest. His two publications for Koishikawa Botanical Garden and the wood panels at Kew, therefore, demonstrate his continued practice of privileging the availability of plants first and foremost.

While it is difficult to ascertain whether the Kew panels were produced from the actual trees grown in Koishikawa Botanical Garden, the direct link between their access and the panels suggests that one of the purposes of these panels was to physically corroborate the breadth of plants growing at Koishikawa Botanical Garden.[58] Here, the demonstrative purpose of the panels comes to the fore. These panels were more than a list of plants: as a set of objects, they embodied their immediate availability in the garden. As a group, they represented a botanical garden in Tokyo.

Tanaka Yoshio's study on wood and taxonomy around this time provides further evidence for the scrutiny of received knowledge within Koishikawa Botanical Garden. In 1872, Tanaka published two charts: *Taxonomy of Plants according to Linnaeus* (Rinnashi shokubutsu kōmōhyō) and *Taxonomy of Plants according to de Candolle* (Dekandorureshi shokubutsu shizen bunkahyō).[59] They were both published by the Museum Bureau (Hakubutsu kyoku), separated by two months.[60] Both charts were foldable to a portable size, suggesting that Tanaka meant them to be taken on excursions, similarly to *Useful Timbers*.

Augustin Pyramus de Candolle's taxonomic classification, published in *Prodromus Systematis Naturalis Regni Vegetabilis* (1824–1872), challenged the established and accepted Linnaean binominal nomenclature based entirely on the numbers of (or lack of) plants' sexual organs – pistil and stamen. Because de Candolle's taxonomy deployed the concept of homology, the number of families, a category one class above genus in the Linnaean system, expanded. This shift in structure brought further complications to the discourse of botanical taxonomy.[61] In fact, Itō was aware of de Candolle's nomenclature, as he referenced it in his 1829 publication, *Nominal Differentiations in Western Materia Medica* (Taisei honzō meiso). Having studied under Itō, Tanaka was not new to the recursive problems of naming and

58 Itō Keisuke begins his "explanatory notes" (*hanrei*) in 1881 *Tokyo Daigaku Koishikawa Shokubutsuen sōmoku zusetsu* by elaborating on the difficulty of names and naming. Of the ten items listed on this note, five delve into issues of names. Itō Keisuke, "*hanrei*," in *Tokyo Daigaku Koishikawa Shokubutsuen*, n.p.

59 Tanaka's translation of de Candolle's system is said to be based on the 1844 edition of *Prodromus Systematis Naturalis Regni Vegetabilis*. Ueno Masuzō, *Nihon hakubutsu gakushi* (Tokyo: Kōdansha, 1989), p. 194.

60 *Rinnashi shokubutsu kōmōhyō* was published in the eighth month, and *Dekandorureshi shokubutsu shizen bunkahyō* in the tenth month. They are both available online at the digital collection of the National Diet Library.

61 The number of families expanded to 213 via de Candolle's system. For an overview of the ramifications of de Candolle's system for the issues of Latin naming, see Dan H. Nicholson, "A History of Botanical Nomenclature," in *Annals of Missouri Botanical Garden*, vol. 78, no.1 (1991): 33–56.

nomenclature. In other words, Tanaka's two taxonomic charts of 1872 served to extend and elaborate on existing nominal concerns that had been shared with figures such as his teacher Itō.

To his own copy of *Taxonomy of Woods according to de Candolle*, now at the National Diet Library in Tokyo, Tanaka added a brief handwritten text explaining his reasons for studying and disseminating this taxonomic system. Perhaps unsurprisingly, here, too, he identified the problem of naming.

> I published a Candollean chart and a Linnaean chart in 1872 [...] The number of available names in classical Chinese is already limited. In recent years, we have been following the names in both classic texts of *materia medica* as well as those published in the Qing dynasty. When a plant lacked a name, we concocted a new translated name [in Japanese]. When the family name was absent in Chinese, we used the name of a plant belonging to that family for the family name. When there were no names available in either classical or Qing texts, we used a Japanese name in this chart.[62]

This note reveals that the matching of the names of plants in various texts and languages continued to be a real challenge for Tanaka and those pursuing knowledge of plants, including trees. Most importantly, Tanaka's effort was directed at aligning various linguistic expressions to facilitate and smoothen the process of identifying his local plants. On this level of methodological concern, Tanaka's motivation matched that of his teacher, Itō.

Just a year after the publication of Tanaka's two charts, the Ministry of Education published *An Introduction to Botany* (Shokugaku senkai). Here, the basic botanical understanding was framed by and through de Candolle's nomenclature. The epistemic negotiations between Candollean and Linnaean systems, therefore, were settled in favour of the former in the context of Meiji education. As a consultant for this publication, Tanaka modelled the Ministry's publication on a textbook by the English botanist John Lindley (1799–1865), *School Botany, and Vegetable Physiology, or The Rudiments of Botanical Science* published in 1860.[63] *An Introduction to Botany* presented Lindley's approach by explaining the fundamental categories based on the shapes of root, stem, bud, leaf, flower, calyx, corolla, stamen, pistil, fruit, and seed. Katō Chikusai was the illustrator for this publication.

62 Augustin Pyramus de Candolle, Tanaka Yoshio trans., *Dekandorureshi shokubutsu shizen bunkahyō* (Tokyo: Monbushō hakubutsukyoku, 1872), https://dl.ndl.go.jp/pid/2543093. Writing in red ink, Tanaka added this text three years after the initial publication.

63 Lindley's textbook is mentioned in the introductory remarks to Ono Motoyoshi trans. *Shokugaku yakusen* (Tokyo: Ministry of Education, 1874), n.p., https://dl.ndl.go.jp/pid/832409.

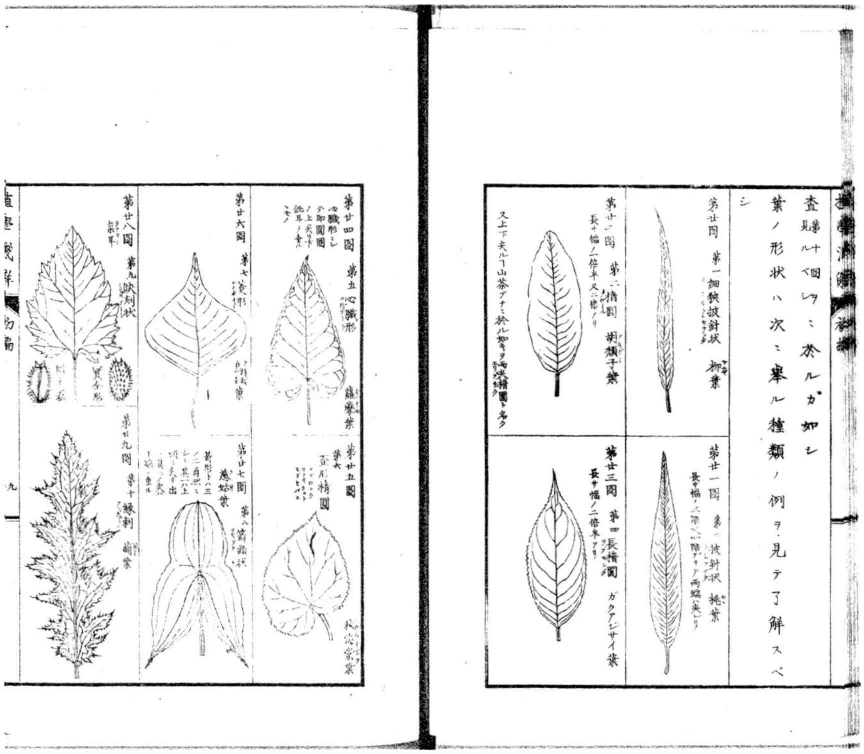

Figure 3.4: A page demonstrating different shapes of leaves in Ono Motoyoshi, Kubo Hiromichi eds., *An Introduction to Botany* (*Shokugaku senkai*), (Tokyo: Ministry of Education, 1875), p. 9.

Tanaka and his team performed a major intervention in this publication. Instead of passively copying the pictorial examples of abstract forms in Lindley's text, Tanaka's team deliberately mobilised specific domestic plants to exemplify the forms by adding the common Japanese names of domestic plants. That is, this publication concretised the categories through specific domestically available plants.[64] For example, in differentiating the shapes of leaves such as oblong and sinuated in Lindley's text, Tanaka's team used the same abstracted pictures of the leaves, asserted common names, lacecap hydrangea (*gaku ajisai*) and cocklebur (*onamomi*), respectively, to demonstrate the differences specifically through their local examples (Figure 3.4).

64 Ono Motoyoshi trans, Kubo Hiromichi ed., and Tanaka Yoshio consultant, *Shokugaku senkai* (Tokyo: Ministry of Education, 1875), n.p. https://dl.ndl.go.jp/pid/832406. When the illustration was a copy from Lindley's book, it was accompanied by a phrase, "in the original." This book was to be paired with the "terms book" or *Shokugaku yakusen* published in 1874.

The introduction to *Useful Timbers* of 1874 made the point that it used de Candolle's taxonomy to frame its body of knowledge and that further information on Candollean nomenclature was available in *An Introduction to Botany*. Here, a recursive and discursive set of texts emerges, in which Itō Keisuke and Tanaka Yoshio had direct leading roles to play. In all publications – *Taxonomy of Plants according to Linnaeus* (1872), *Taxonomy of Plants according to de Candolle* (1872), *An Introduction to Botany* (1873), *Useful Timbers* (1874), *Catalogue of Plants in Koishikawa Botanical Garden* (1877), and *Figures and Descriptions of Plants in Koishikawa Botanical Garden* (1881) – a decisive gesture was made to establish a direct and firm relationship of fidelity among the locally available trees and their various names. The relationship had economic, botanical, and educational ramifications. In the case of *Useful Timbers,* the inclusion of physical pieces of timber was intended to encourage readers to bridge the gap between the actual structure and the represented knowledge on the page. At the same time, the deployment of de Candolle's nomenclature in this book signalled, for the Ministry of Education, and Itō and Tanaka especially, an epistemic experiment.

As if to further their exploration, the Ministry of Education published a series of pedagogical charts to be used in elementary schools a year later, in 1876. One of them is titled *Hakubutsufu daiichizu* or *Chart of Plants 1*, and here, too, de Candolle's nomenclature is illustrated using local examples (Figures 3.5 and 3.6). Printed in copper-etching and coloured in woodblock, the chart names Katō as illustrator.[65] Katō evidently traced and reused illustrations from *An Introduction to Botany* for this coloured chart (Figure 3.4). These repeated efforts to provide local examples and names to explain de Candolle's nomenclature bring another different dimension to the idea of usefulness.

For example, in the Linnaean nomenclature, according to Itō, the tree commonly known in Japan as *sugi* is rendered *Cupressus Japonica* Linn. In de Candolle's nomenclature, according to Tanaka's 1872 chart, it is rendered *Cryptomeria Japonica* due to further differentiation in the category of family. On the Kew panel, it is identified as *Cryptomeria japonica*, Don. A common tree known as *kusunoki* in Japanese was identified with two binominal names by Itō in 1829: *Laurus Champhora* Linn: and *Cinnamomum Camphora* Sieb. In Tanaka's *Useful Timbers* it is noted as *Cinnamomum Camphora* Fr. On the Kew panel, its Latin name is rendered *Cinnamomum camphora*.

Nominal Differentiations in Western Materia Medica (1829), *Useful Timbers* (1874), and the twenty-six wood panels at Kew have twenty trees in common. All the names of these twenty trees are rendered differently in Itō's 1829 *Nominal Differentiations*

65 The format of the charts is said to be based on the series published by N. Wilson and N.A. Kalkin's *School and Family Charts*.

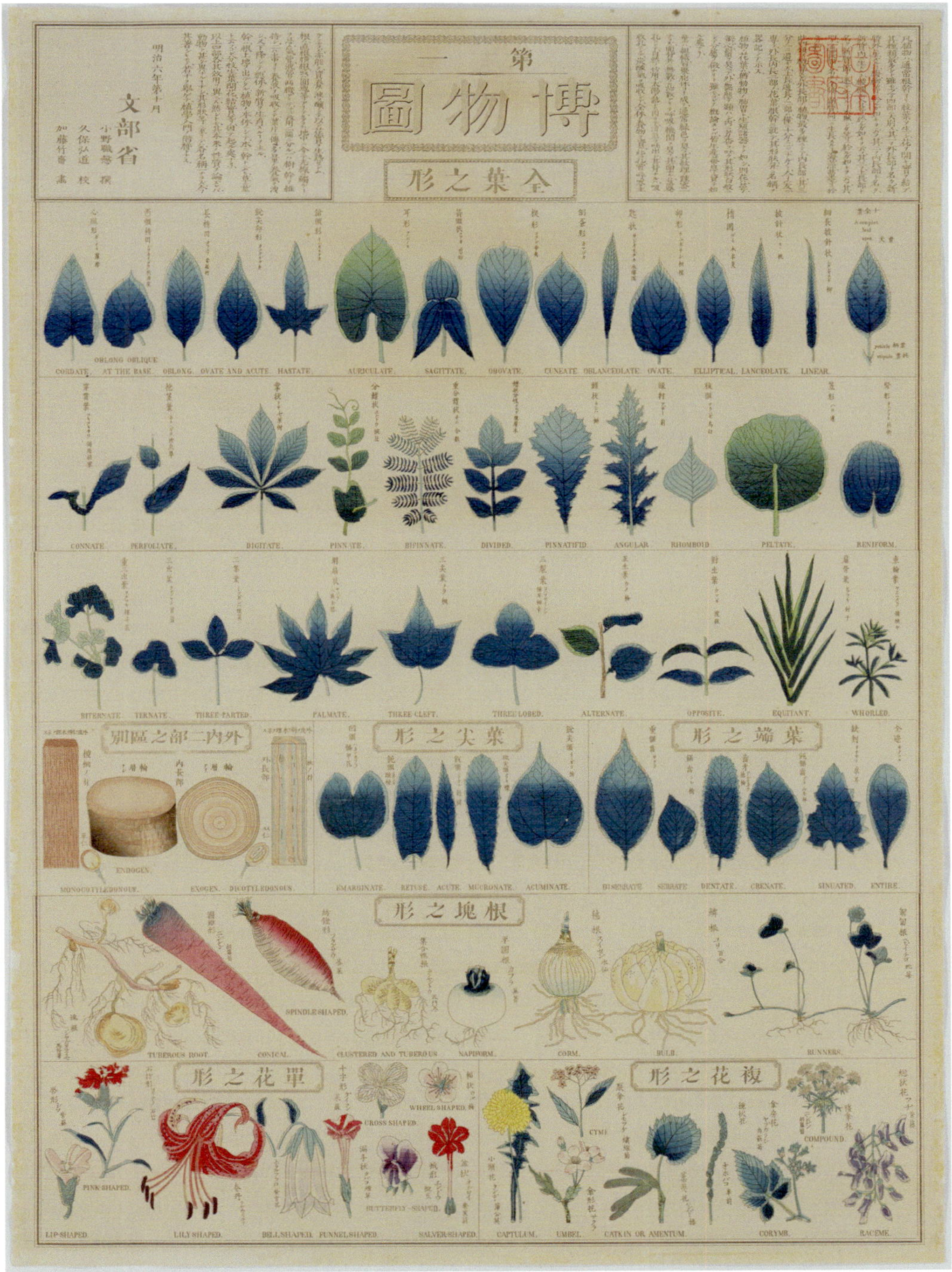

Figure 3.5: Ministry of Education, *Chart of Plants 1* (*Hakubutsufu daiichizu*) 1876. copper-printing with woodblock colours. 58 cm x 81 cm.

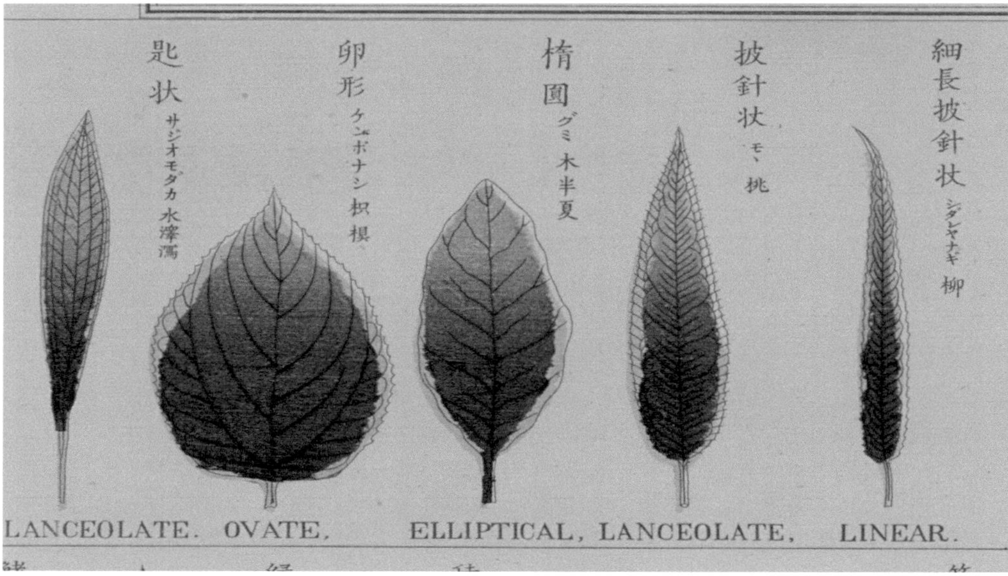

Figure 3.6: Close-up of Figure 3.5, *Chart of Plants 1 (Hakubutsufu daiichizu)* 1876.

in Western Materia Medica and Tanaka's 1874 *Useful Timber*. Here, we come to see how the stakes of knowledge production and application played out through and around the wood panels. They do not merely function as a confirmatory device of the trees' existence within the Botanical Garden, but likewise serve to discuss and correct the given binominal names. Against the background of the government's ideas about the export of timber products as a nation-building project, Tanaka and Itō were also faced with a different challenge, shaped by their desire to align and consolidate various nomenclatures with the available trees in Tokyo, in their Botanical Garden.

Such historical probing and intellectual investment are overlooked when we blindly rely on the putative idea of *botany* and see these objects as supporting evidence for their "accurate" transfer to Japan. As carriers of a particular body of knowledge consisting of physical parts, corresponding pictorial elements, and the assigned names, these panels served as intermediaries to verify nomenclature. The insights gained from their creation in turn contributed to the further production of knowledge. Here, the issue is neither "enlightenment" or "civilisation," as imagined by Fukuzawa and Meiji politicians who developed his ideas in practice, nor economic interest. Naming, un-naming, and re-naming, a familiar approach to the study of plants, involved Tanaka and Itō in an ongoing epistemic process. The Koishikawa Botanical Garden served as a laboratory for never-ending experiments with nomenclature and taxonomy.

What is Inventive About the Wood Panels?

What, then, can we make of Katō's red stamp that declared "1878, the latest invention by Katō Chikusai"? In the context of the emerging political and economic interest in identifying and profiting from natural resources, the Kew panels reveal themselves as a set of objects that served several different purposes.

In their portable and stackable format, the panels can be collected as a reference guide with samples of "useful plants." When situated in the context of international and domestic expositions, the panels as a group materially represent the variety of trees in Koishikawa Botanical Garden. The Exhibition Bureau saw the panels as a commodity for the education of the general public. Compared to *Useful Timbers*, which took the equally portable and stackable form of the published book, the wood panels were able to carry –literally and materially – more information such as the textures of the bark and the illustrations of its flowers and fruits. The wood panels also served as a mediating device with which the ascribed binominal names could be challenged, modified, or at least discussed by giving immediate and direct access to the visual appearance, scent, and texture of the tree in question.

In comparison to *Useful Timbers,* the newness suggested by the phrase "the latest invention" gestures towards an improved and more efficient format to present a body of knowledge. Given the collaboration of the Museum Bureau, Exhibition Bureau, and the Koishikawa Botanical Garden, and the personal relationship between Tanaka Yoshio and Itō Keisuke, *Useful Timbers* and the panels share the intention to educate the public and to create a modality of learning to acquaint it with the physical aspect of a tree. Juxtaposing these two formats of knowledge transfer, it is possible to view the panels as an improved version of *Useful Timbers*. While it is speculative, given the fact that Tanaka participated in the governmental discussions on the issue of patent law, it would not be surprising if he regarded the panels as a potential case study to concretise matters.

At the same time, from the Koishikawa Botanical Garden perspective, the panels served to address a different set of concerns. For Itō and Tanaka, they were an efficient device to display and discuss Latin names, and, as such, offered a valuable tool in the ongoing process of refining nomenclature, even more so because they focused on plants whose Latin names had changed since Itō's 1829 publication.

The fact that Katō Chikusai supplied illustrations for *An Introduction to Botany, Figures and Descriptions of Plants in Koishikawa Botanical Garden*, and the elementary school chart, *An introduction to botany* as well as for the panels also points to the extent of Katō's own knowledge about trees and botanical nomenclature. The novelty of Katō's "invention" was the combination of text (in various forms and scripts), image, and sample – a format that departed from the pictorial work he had produced for Itō and Tanaka.

As a way of concluding, let us return to the disciplinarity of the Kew panels as objects. Upon considering motivations in production, a sense of unruliness emerges from the co-existence of multiple values within and around these objects. As a group of objects, they disrupt the methodological assumptions that constitute modern subject areas, such as botany and art history. They ask us to suspend our habits of disciplinary contextualisation. On one level, the objects' value is in their function as a teaching tool: they served as a conduit to transfer a particular kind of knowledge. On another level, they helped to identify useful and therefore lucrative natural resources. On yet another level, the same objects also benefited the lives of Itō and Katō, both financially and intellectually. Acting in defiance of modern disciplines, boundaries, and definitions, these objects are ill-disciplined.

Glossary

Katō Chikusai (1818–1886)	加藤竹斎
Itō Keisuke (1803–1901)	伊藤圭介
Koishikawa	小石川
materia medica (J: *honzō*, Ch: *bencao*)	本草
Hakubutsufu daiichizu	博物譜第一図
Honzōgaku	本草学
Shokubutsugaku	植物学
Takahashi Korekiyo (1854–1936)	高橋是清
tokkyokyoku	特許局
hatsumei	発明
hakken	発見
Tanaka Yoshio (1838–1916)	田中芳男
Yūyōmokuzai shōran	有用木材捷覧
Shokugaku senkai	植学浅解

Bibliography

Primary Sources

Abe Rekisai 阿部櫟斎. *Sōmoku sodategusa* 草木育種. Tokyo: Inada Sahee, 1876.

Dainihon sanrinkai 大日本山林会ed. *Tanaka Yoshiokun nanaroku tenrankai kinenshi* 田中芳男君七六展覧会記念誌. Tokyo: Dainihon sanrinkai, 1926.

Dainihon yōgyōkyōkai 大日本窯業協会. *Nihon kinsei yōgyōshi* 日本近世窯業史. Tokyo: Dainihon yōgyōkyōkai, 1922.

De Candolle, Augustin Pyramus 垤甘度爾列. Tanaka Yoshio 田中芳男trans., *Dekandorureshi shokubutsu shizen bunkahyō* 垤甘度爾列氏植物自然分科表. Tokyo: Monbushō hakubutsukyoku, 1872.

Fujiwara Masato 藤原正人. Ed. *Meiji zenki sangyō hattatsushi shiryō* 明治前期産業発達史資料. Tokyo: Meiji bunken shiryō kankōkai, 1963.

Fukuzawa Yukichi 福澤諭吉. *Seiyo jijō* 西洋事情. Kyoto: Hayashi Yoshibei, 1868.

Genrōin 元老院. *Genrōin kaigihikki* 元老院會議筆記. Tokyo: Genrōin, 1880.

Hakurankai jimukyoku 博覧会事務局, *Yūyō mokuzai shoran* 有用木材捷覧. Tokyo: Hakurankai jimukyoku, 1874.

Hepburn, James Curtis. *A Japanese and English Dictionary*. Shanghai: American Presbyterian Mission Press, 1867.

Itō Keisuke, *Catalogue of Plants in Koishikawa Botanical Garden*. Tokyo: University of Tokyo, 1879.

Kaneko Kōichirō 金子孝一郎. *Senbaitokkyojōrei chūshaku* 専売特許条例註釈. Niigata: Miyajidō, 1885.

Maeda Kenkichi 前田献吉. *Nihon yūyōmokuzai hyōhon kaisetsu* 日本有用木材標品解說. Tokyo: self-published, 1889.

Miyata Uhei 宮田宇兵衛. *Tokyo shotaika raimeikyō* 東京諸大家雷銘鏡. Tokyo: Miyata Uhei, 1880.

Morse, Edward S. *Japan Day by Day*. New York: Houghton Mifflin Company, 1917.

Naikaku kanpōkyoku 内閣官報局. *Hōrei zensho* 法令全書. Tokyo: Naikaku kanpōkyoku, 1912.

Ono Motoyoshi 小野職愨trans. *Shokugaku yakusen* 植学訳筌. Tokyo: Ministry of Education, 1874.

Ono Motoyoshi 小野職愨, Kubo Hiromichi 久保弘道 ed., and Tanaka Yoshio 田中芳男 consult, *Shokugaku senkai* 植學淺解. Tokyo: Ministry of Education, 1875.

Otsuki Kinju 大槻勤寿, ed. *Meiji jūgonen kaiga kyōshinkai shuppin gaka jinmei ichiran* 明治一五年絵画共進会出品画家人名一覧. Osaka: Otsuki Kinju, 1883.

Smiles, Samuel. Translated by Nakamura, Masao. *Saigoku risshihen: genmei jijoron*. Tokyo: Suharaya Mohei, 1870.

Takahashi Korekiyo 高橋是清. *Takahashi Korekiyo jiden* 高橋是清自伝. Tokyo: Chikura shobō, 1936.

Takahashi Korekiyo 高橋是清. "Wagakuni tokkyoseido no kiin 我が国特許制度の起因." https://www.iip.or.jp/chizaishi/pdf/korekiyo2.pdf

Takamura Kōun 高村光雲. *Bakumatsu ishin kaikodan* 幕末維新懷古談. Tokyo: Iwanami shoten, 1995.

Tanaka Yoshio 田中芳男, Hirayama Narinobu 平山成信. *Ōkokuhakurankai sandō kiyō* 澳国博覧会参同記要. Tokyo: Moriyama Shunyō, 1897.

Tokio Daigaku, *Catalogue of Plants in Koishikawa Botanical Garden*. Tokyo: Scientific Department of Tokio Daigaku, 1877.

Tōkō Gakujin 東江学人. *Bunmei kaika naigai jijō* 文明開化内外事情. Tokyo: Higashinari Kamejirō, 1873.

Tokyo kokuritsu bunkazai kenkyūjo 東京国立文化財研究所, *Naikoku kangyō hakurankai bijutsuhin mokuroku* 内国勧業博覧会美術品出品目録. Tokyo: Tokyo kokuritsu bunkazai kenkyūjo, 1996.

Wagner, Gottfried. Translated by Asami Tadamasa 浅見忠雅. *Meiji jūnen naikoku kangyō hakurankai hōkokusho* 明治十年内国勧業博覧会報告書. Tokyo: Naikoku kangyō hakurankai, 1877.

Secondary Sources

Dear, Peter. *The Intelligibility of Nature: How Science Makes Sense of the World.* Chicago, IL: The University of Chicago Press, 2006.

Fukuoka Maki. *The Premise of Fidelity: Science, Visuality and Representing the Real in Early Nineteenth-Century Japan.* Stanford, CA: Stanford University Press, 2012.

Heath, Christopher. "Intellectual Property and Anti-Trust," in Wilhelm Röhl ed., *History of Law in Japan Since 1868*. Leiden: Brill, 2005, pp. 402–543.

Itō Mamiko 伊藤真実子. *Meiji nihon to bankoku hakurankai* 明治日本と万国博覧会. Tokyo: Yoshikawa kōbunkan, 2008.

Kinoshita Naoyuki 木下直之. *Bijutsu toiu misemono* 美術という見世物. Tokyo: Heibonsha, 1993.

Kinoshita Naoyuki 木下直之. *Shashingaron: shashin to kaiga no kekkon* 写真画論: 写真と絵画の結婚. Tokyo: Iwanami shoten, 1996.

Kinoshita Naoyuki 木下直之. "Daigaku nankō bussankai nitsuite 大学南校の物産会について," in University of Tokyo ed., *Gakumon no arukeorojī* 学問のアルケオロジー. Tokyo: University of Tokyo Press, 1997, pp. 86–104.

Kinoshita Naoyuki 木下直之. *Yo no tochū kara kakusareteiru koto—kindai nihon no kioku* 世の途中から隠されていること: 近代日本の記憶. Tokyo: Shōbunsha, 2002.

Kobayashi Satomi 小林聡. "Edo jidai ni okeru hatsumei sōsaku to kenrihogo 江戸時代における発明・創作と権利保護," *Patent* パテント 61.5 (2008): 48–55.

Komonroku, vol. 96 (April 1885), Ministry of Agriculture and Commerce, n.p.

Lack, H. Walter. "Plant Illustration on Wood Blocks: A Magnificent Japanese Xylotheque of The Early Meiji Period," *Curtis' Botanical Magazine* 16.2 (1999): 124–134.

Lewis, Martin W. and Kären Wigen. *The Myth of Continents: A Critique of Metageography.* Chicago, IL: University of Chicago Press, 1997.

Murata, Hiroko 邑田裕子, Sakazaki Nobuyuki 坂﨑信之, Sakurada Michio 桜田通雄, Tōma Tetsuo 東馬哲雄, and Murata Jin 邑田 仁. "'Meiji jūshichinen jūnigatsu Tokyo Daigaku shokubutsuen shaseizu (mokuroku)' oyobi Koishikawa Shokubutsuen shōzō mokuzai hengaku ni tsuite 明治十七年十二月東京大学植物園写生図」(目録)および小石川植物園所蔵木材扁額について," in *Itō Keisuke nikki* 伊藤圭介日記. Volume 19. Nagoya: Higashiyama Botanical Garden, 2014, pp. 203–226.

Nagata Toshiyuki, Ashley DuVal, Hans Walter Lack, George Loudon, Mark Nesbitt, Michaela Schmull, and Peter Crane. "An Unusual Xylotheque with Plant Illustrations from Early Meiji Japan," *Economic Botany* 67.2 (2013): 87–97.

Nagata Toshiyuki 長田敏行. "Shokubutsugahō sōmeiki no tsuiseki – Katō Chikusai hengaku kara 植物画 植物画法黎明期の追跡 —加藤竹斎扁額から," *Seibutsu no kagaku iden* 生物の科学遺伝 71.3 (May 2017): 190–194.

Nappi, Carla. *The Monkey and the Inkpot: Natural History and Its Transformations in Early Modern China.* Cambridge, MA: Harvard University Press, 2009.

Nappi, Carla. "The Global and Beyond: Adventures in the Local Historiographies of Science," *ISIS* 104 (2013): 102–110.

Nesbitt, Mark. "Puzzling Panels," *Kew Magazine* (Spring 2010): 54–55.

Nicholson, Dan H. "A History of Botanical Nomenclature," *Annals of Missouri Botanical Garden* 78.1 (1991): 33–56.

Norota Junichi 野呂田純一. *Bakumatsu Meiji no biishiki to bijutsuseisaku* 幕末・明治の美意識と美術政策. Kyoto: Miyaobi shuppansha, 2015.

Oba Hideaki 大場秀章 ed. *Nihon shokubutsukenkyū no rekishi: Koishikawashokubutsuen sanbyakunen no ayumi* 日本植物研究の歴史: 小石川植物園300年の歩み. Tokyo: University of Tokyo Museum, 1996.

Röhl, Wilhelm, ed. *History of Law in Japan Since 1868.* Leiden: Brill, 2005.

Satō, Dōshin. *Modern Japanese Art and the Meiji State: The Politics of Beauty.* Los Angeles, CA: Getty Publications, 2003.

Seki Hideo 関秀夫. *Hakubutsukan no tanjō: Machida Hisanari to Tokyo Teishitsu hakubutsukan* 博物館の誕生: 町田久成と東京帝室博物館. Tokyo: Iwanami shoten, 2005.

Shiina Noritaka 椎名仙卓. *Meiji hakubutsukan kotohajime* 明治博物館事始め. Tokyo: Shibunkaku shuppansha, 1989.

Sugimoto Isao 杉本勲. *Itō Keisuke* 伊藤圭介. Tokyo: Yoshikawa kōbunkan, 1960.

Suzuki Hiroyuki 鈴木廣之. *Kōkokatachi no jūkyūseiki: Bakumatsu Meiji ni okeru monono arukeoroji* 好古家たちの19世紀: 幕末明治における《物》のアルケオロジー. Tokyo: Yoshikawa Kōbunkan, 2003.

Takagi Hiroshi 高木博志. "Nihon bijutushi no seiritsu shiron 日本美術史の成立・試論," *Nihonshi kenkyū* 日本史研究 400 (1995): 74–98.

Tseng, Alice. *The Imperial Museums of Meiji Japan.* Seattle, WA: University of Washington Press, 2008.

Ueno Masuzō 上野益三. *Nihon hakubutsu gakushi* 日本博物学史. Tokyo: Kōdansha, 1989.

University of Tokyo 東京大学ed., *Tokyo Daigaku Koishikawa shokubutsuen sōmoku zusetsu* 東京大学小石川植物園草木図説. Tokyo: Maruzen, 1886.

Yoshimi Shunya 吉見俊哉. *Hakurankai no seijigaku: manazashi no kindai* 博覧会の政治学: まなざしの近代. Tokyo: Chūōkōron, 1992.

About the Author

Maki Fukuoka is an art historian whose research and teaching interests are coordinated by two broad axes: histories of seeing and modes of knowing. Her first book, *The Premise of Fidelity. Science, Visuality, and Representing the Real in Nineteenth-century Japan* situates histories of seeing in Tokugawa Japan and their relationship to an alternative mode of knowing in the context of medical science. She has published several articles and chapters on the visual and material culture of nineteenth-century Japan. In 2022, the Getty Institute published her translation of Suzuki Hiroyuki's book (Japanese to English) as *Antiquarians of Nineteenth-century Japan. The Archaeology of Things in Late Tokugawa and Early Meiji Periods.* Since 2016, she has also been engaged with a collaborative project to articulate and challenge the relationship between gender issues and contemporary art in East Asia through the project Engendering East Asia (https.//engenderingeastasia.org/).

4. The Return of the Elephants: A Social History of Elephant Watching in Early Modern China

Fan Lin

Abstract: This chapter examines the social history of watching elephants in late imperial China. Elephants were native to early China, but they had migrated to the southwestern margin of the country and Southeast Asia by the tenth century. After the Song dynasty, they were brought back to the capital as tributary animals or diplomatic gifts. They entered the public arena by, for example, being used in imperial parades. In the Ming dynasty, the washing of elephants in the city moat outside the imperial palace in early summer evolved into a captivating urban spectacle. This phenomenon instigated the creation of various artworks but also took on additional layers of social significance.

Keywords: Elephant, *xixiang tu* 洗象, xiangfang 象房

While multiple philosophical discourses in East Asia highlighted the concept of nature (*ziran*, *shizan*), the societies that gave rise to these ideas were predominantly human-centred. The significance of animals, their roles, functions, and symbolic meanings were largely defined by their interaction with human society. The elephants, the largest terrestrial creatures today, serve as an illustration of how animals have been treated: they were ascribed auspicious and noble meanings on one hand; yet, on the other, they were subjected to ruthless exploitation as useful resources. Informed by Wen Huanran's research, Mark Elvin directs the attention of Western readers towards the enduring conflict between elephants and humans throughout the history of China.[1] "Retreat" in the title of his book, *The Retreat of the Elephants: An Environmental History of China*, captures the ill-fated destiny of wild elephants, which migrated to the southernmost reaches of the continent in the tenth century due to climate change and, more importantly, the expansion of human

1 Wen Huanran et al., *Zhongguo lishi shiqi zhiwu yu dongwu bianqian yanjiu* (Chongqing: Chongqing chubanse, 1995), pp. 186–215. Mark Elvin, *The Retreat of the Elephants: An Environmental History of China* (New Haven, CT: Yale University Press, 2004), pp. 9–18.

Fan Lin and Doreen Mueller (eds), *Picturing Animals and Plants in Early Modern China and Japan. Innovation, Experiments, and Anxieties*. Taylor & Francis Group, 2025
DOI: 10.5117/9789048559091_CH04

activities. At the end of the first millennium, human interaction with elephants shifted to trained ones. In the coming centuries, these creatures were transported from Southeast Asia and southwest China, predominantly to the capitals in north China as diplomatic gifts as well as tributary animals.

Mainly used in court rituals, such as imperial processions and daily morning courts, elephants returned to the capitals as exotic, regal animals as an urban spectacle and generated new cultural meanings. Recent research by Cheng Minsheng, Liu Xiangxue, and Hui-Chun Yu has revealed interests in the provenance of these elephants and their institutional roles in imperial rituals from the Song (960-1279) to the Qing (1644–1911) periods.[2] Elephant performances came to be institutionalised in the Song dynasty in the sense that their acquisition, domestication, and training were made possible under recently installed policies and facilities. The underlying logic of keeping elephants for court rituals was maintained until the end of the Qing dynasty. Hui-Chun Yu argues that elephants were imperialised by the palace machine, which "provided the framework and incentive to acquire, possess, domesticate, train, transform, rank and integrate them into the political performances of the Qing monarchy."[3]

This chapter mainly focuses on a different side of the story of the 'return' of the elephants, that is, elephant watching. In social reality, elephants never existed on their own; they were often targets of urban gaze. Although elephant keeping and training were part of the imperial machine, the intended grandeur of their performance could only be realised while they were being watched by the populace. During this process, however, elephant watching gradually shifted away from the imperial vision and was woven into the fabric of urban life. This chapter first problematises the timeless symbolism of elephants as auspicious animals and then examines the connotations of elephants in cosmic-political rhetoric. The second part investigates how the image of elephants as 'political animals' was received among urbanites in the capitals. When the elephants on the training ground started to ignite public interest, the meaning of elephants was given an innovative twist towards an entertaining dimension. The third part examines a special phenomenon, elephant bathing. It examines why, although elephants, as performers of court rituals, had to be bathed regularly, their bathing only became an urban spectacle in the Ming (1368-1644). The last part of this chapter examines the relationship between elephant bathing as an urban event and as a theme of Buddhist paintings. From a visual culture

2 Liu Xiangxue, "Mingdai xunxiangwei kaolun," *Lishi yanjiu* (2011.1): 51–66; Cheng Minsheng, "Songdai daxiang de ziran yu shehui shengtai," *Zhongyuan wenhua yanjiu* (2021:3): 56; Hui-Chun Yu, "When There Is Peace, There Are Elephants," in Marina Siebert, Chen Kai Jun, and Dorothy Ko, eds., *Making the Palace Machine Work: Mobilizing People, Objects, and Nature in the Qing Empire* (Amsterdam: Amsterdam University Press, 2021), pp. 271–290.

3 Yu, "When There Is Peace, There Are Elephants," p. 272.

perspective, during the *longue durée* from the Song to the late Ming, when Chinese empires witnessed gradual urbanisation and vernacularisation of court practices, it was the watching and bathing of elephants in public spaces that removed elephants from the political arena and created multifaceted, interconnected meanings of their existence in popular culture in early modern China.

New Concepts: Wild vs. Tamed Elephants

The Northern Song dynasty (960-1127) saw a critical phase in the trajectory of the elephants' fate. In the first decade of the dynasty, wild elephants were spotted for the last time in central China and the mid-south (present-day Henan, Hubei, and Hunan provinces). This was the time when the Chinese climate started to cool down and urbanisation gained momentum. The temperature on the East Asian continent started to drop systematically from the eleventh century, prompting profound ecological changes as well as migration of humans and animals to the south.[4] In addition, expanded human settlements encroached into the habitat of elephants that usually travelled long distances in search of food.[5] Although elephants in early and medieval China were symbolically auspicious, wild elephants were treated in a different category. In many cases, their visits to human settlements were unwelcome. In 962, it was reported that an elephant wandered into the woods in Huangpi county (in Hubei province) and ruined people's crops. After it had travelled a surprisingly long distance, from An prefecture to Fu, Xiang (in Hubei province) and Tang (in Henan province) prefectures, it was finally captured at the end of 963 and executed in Nanyang (in Henan province), with its tusks and skin being removed and submitted to the court.[6] Alongside other entries about elephants, the above-mentioned anecdote is recorded as an abnormal phenomenon of wild animals in the 'Treatise on the Five Phases' (Wuxing zhi) in *Song History* (Song shi). Since the Han dynasty (206 BCE–220 CE), the 'Treatise on the Five Phases' had been the official interpretation of how all abnormal, mostly harmful phenomena could reflect the gains and losses of the rule of government and court. Therefore, in a political-cosmic reading of these events, most of these wild elephants, just like other wild animals such as tigers, symbolised destructive potency, posing threats to the proper order of the cosmos.

By contrast, trained elephants presented a different image. Just like other tributary animals, a trained elephant was a sign of submission to the authority of

4 Zhu Kezhen, *Tiandao yu renwen* (Beijing: Beijing chubanshe, 2006), pp. 77–94.

5 Wen Huanran, "Zaitan lishi shiqi de yexiang fenbu," *Zhongguo lishi shiqi zhiwu yu dongwu bianqian yanjiu*, pp. 205–206.

6 Tuotuo, *Song shi* (Beijing: Zhonghua shuju, 1977), 66/1450.

the central court. In 961, three tributary animals, namely, a yellow parrot, a tamed elephant, and a white hare, appeared on the banner used for the Southern Suburban Sacrifice, the most significant court ritual that offered sacrifice to Heaven once every three years.[7] In the third month in 968, a tamed elephant was recorded as "having arrived in the capital by itself" (*zi zhi jingshi*). The officials immediately composed memorials in praise of this event.[8] In the sixth month, four musical episodes on themes of the elephant, a piece of auspicious wood, a white horse, and a white sparrow, were presented at the suburban sacrifice.[9] The eulogy for the episode on the elephant reads as follows,

> Magnificent, that tamed elephant!
> Submitting itself to the realm of the emperor,
> Its nature cultivated in the southern regions,
> It offers auspicious service to the central area.
> The emperor's benevolence called upon the giant beast to arrive,
> The emperor's virtue placated the distant and the wild.
> The elephant was moved by this resonance,
> And its miraculous transformation is boundless.[10]

Here, the tamed, tributary elephant was staged as an auspicious animal paying respect to far-reaching imperial power. In the early Northern Song, although the court repeatedly issued orders forbidding the capture of wild elephants in the Lingnan region (in present-day Guangdong and Guangxi), capturing wild elephants, which were deemed life-threatening, was still a common practice in central and mid-south China.[11] It was not until 967 that an Office for the Care of Elephants (Yangxiang suo) was established after a contingent event. In the eighth month of that year, ten days after a wild elephant meandered into the capital, the court allegedly sent five hundred soldiers to capture it and subsequently established the office.[12] By the end of the first decade of the Northern Song, tamed elephants serving as exotic animals in court activities had replaced the image of wild elephants in public memory.

Cheng Minsheng makes the shrewd observation that, in Tang China (618–906), elephants were kept together with horses and oxen in imperial stables and trained for

7 Zhao Sheng, *Chaoye leiyao* (Zhi bushuzhai edition, 1813), 1.3b.

8 Li Tao, *Xu zizhi tongjian changbian* (Beijing: Zhonghua shuju, 2004), vol. 1, 9/201.

9 Xu Song, "Yangxiang suo," in *Song Huiyao jigao* (Taipei: Academia Sinica, Chengdu: Sichuan daxue, Cambridge, MA: Harvard University, 2008), Zhiguan 23.3.

10 Tuotuo, *Song shi*, 134/3130.

11 It is worth noting that government measures of prohibiting the killing of wild elephants in the Lingnan region was made exactly because this was happening on the ground. Xu Song, "Daojia xiang," in *Song Huiyao jigao*, in Yufu 3.15. Also See Cheng Mingsheng, "Songdai daxiang de ziran yu shehui shengtai."

12 Xu Song, "Yangxiang suo," in *Song Huiyao jigao*, Zhiguan 23.3.

musical performances but were not considered a distinct category.[13] It was only during the Song that elephant domestication and training were systematically carried out. The above-mentioned Office for the Care of Elephants was built in Yujin Garden, the imperial garden outside Nanxun palace gate in the capital. The office soon received more elephants within and outside the Song empire.[14] Eventually, the number of the elephants reached forty-five. Like soldiers in processions, elephants were also trained to follow orders, and banners were made for this purpose. A piece of land measuring fifteen *qing* (about one square kilometre) in Yujin Garden provided hay to feed the elephants.[15] Given the fact that an elephant would consume 150 kilograms of food and forty litres of water a day, it was impossible for the city to keep this number of elephants at this venue throughout the year. Therefore, to herd the animals from the fourth to the ninth month, a second elephant stable was built one hundred kilometres away from the capital, near Bian River in Ningling county (in Henan province).[16]

The practice of using elephants in court rituals continued in later dynasties. After the Yuan court (1279–1368) established itself in Beijing, it continued with the practice of keeping elephants. The first elephant stable was built to the north of Haizi Bridge, a popular location connecting the Grand Canal and the lake of Jishuitan. Scholar Xiong Mengxiang (fourteenth century) witnessed the arrival of an elephant to the stable on the first day of the first month in 1297. This elephant was said to be taller than normal elephants and moved as fast as a horse. Given the location of the bridge and the date, it is possible that the event was planned to create a festive effect. In the late Yuan, the elephant stables were moved to Qincheng, a Beijing suburb, where hot springs were available for their bathing.[17]

In the Ming dynasty, a considerable number of elephants were used in the morning court service and imperial procession. In addition to Guangdong and Guangxi, trained elephants were also imported from Yunnan and other tributary states including Annam, Champa, Chenla, Siam, and Myanmar.[18] The increase in the number of

13 Cheng Mingsheng, "Songdai daxiang de ziran yu shehui shengtai," p. 56, endnote 3. Liu Xu, *Jiu Tang shu* (Beijing: Zhonghua shuju, 1975), 7/140, 8/166, 28/1058; Li Longji comp., *Tang liu tian*, in *Yingyin Wenyuange Siku quanshu* (Taipei: Taiwan shangwu yinshuguan, 1983), vol. 595, 17.11b-12a (171). Cheng Minsheng, "Songdai daxiang de ziran yu shehui shengtai," *Zhongyuan wenhua yanjiu* (2021:3): 56.

14 Take the records in the "Fanyi" section in *Song Huiyao jigao* as an example, the Song court received elephants from Guangxi in 1051, 1055 (5.61, 7.29), Wuyue (907-978) in 976 (7.6), Đại Việt (present-day northern Vietnam) in 998, 1001, 1046–1047, 1055–1056, 1063, 1146, 1156, 1161, 1174 (4.25, 4.33–34, 4.44, 4.47, 7.13–7.14, 7.27, 7.30,7.46, 7.47, 7.49, 4.55), Champa (present-day central and southern Vietnam) in 983, 1042, 1061 (4.64, 4.70–71, 7.26, 7.30), Chenla (present-day Cambodia) in 1155 (7.47), Lavo (present-day Thailand) in 1155 (7.47).

15 Xu Song, "Yangxiang suo," in *Song Huiyao jigao*, Zhiguan 23.3.

16 Xu Song, "Yangxiang suo," in *Song Huiyao jigao*, Zhiguan 23.3.

17 Xiong Mengxiang, *Xi jin zhi jiyi* (Beijing: Beijing guji chubanshe, 1983), pp. 212, 232.

18 Huang Yong, "Mingdai jingshi chaoxiang laiyuan kao," *Gugong xuekan* (2008): 159–171. Liu Xiangxue, "Mingdai xunxiangwei kaolun," *Lishi yanjiu* (2011.1): 58.

elephants was accompanied by further regulations concerning elephant keeping. During the Hongwu Era (1368–1398), an Elephant Training Guard (Xunxiang wei) was established in Siming prefecture and later moved to Hengzhou prefecture (both in Guangxi).[19] As it functioned on the frontier of the empire, its purpose was not only to procure elephants but also to guard the territory. The so-called guard (*wei*) was the main military operation unit in the Ming dynasty. A guard usually comprised of 5,600 soldiers under the command of a guard commander (*zhihui shi*).[20] Other peripheral regions likewise presented elephants. In the year 1388 alone, Guangdong presented 123 elephants;[21] after the conquest of the Luchuan headmen (*tusi*), thirty-seven elephants were presented to the capital, and the captives were kept as mahouts at the Elephant Training Guard;[22] Yunnan presented five hundred elephants and three hundred mahouts;[23] Siam presented thirty elephants;[24] Chenla presented twenty-eight elephants as well as thirty-four mahouts;[25] Taiping and Siming prefectures (in Guangxi province) also presented unknown number of elephants.[26]

During the Ming, the captured elephants were first transported to the Elephant Training Office (Xunxiang suo). This was a unit of the Embroidered-Uniform Guard (Jinyi wei), where mahouts trained elephants for the purpose of court audiences, drawing carriages, and carrying treasures.[27] After the Ming capital was moved to Beijing, elephants were usually kept at the elephant stable (xiangfang), which was located at the Shooting Battalion (She suo) on western Chang'an Street.[28] Elephant stables were first built on the old Yuan dynasty site near Haizi Bridge and in 1495 new stables were built near Xuanwu Gate of the inner city (Figure 4.1a).[29]

Elephant Watching

From the Northern Song onwards, a certain number of elephants were always kept in capitals, where they could easily be accessible for ritual performance. *Records of the Prosperity and Scenic Sites* (Fansheng lu) by the Elder from the West Lake

19 Liu Xiangxue, "Mingdai xunxiangwei kaolun," 54.

20 Michael Szonyi, *The Art of Being Governed: Everyday Politics in Late Imperial China* (Princeton, NJ: Princeton University Press, 2017), pp. 53–54.

21 *Ming shilu* (Taipei: Zhongyang yanjiuyuan, 1963), Taizu/188/2816-17.

22 *Ming shilu*, Taizu/189/2860, Taizu/192/2886.

23 *Ming shilu*, Taizu/190/2868.

24 *Ming shilu*, Taizu/193/2893.

25 *Ming shilu*, Taizu/193/2904.

26 *Ming shilu*, Taizu/193/2905.

27 Zhang Tingyu, *Ming shi* (Beijing: Zhonghua shuju, 1974), 76/1862.

28 Shen Defu, *Wanli yehuo bian* (Beijing: Zhonghua shuju, 1997), p. 607.

29 Shen Shixing, *Da Ming huidian* (Taipei: Guofeng chubanshe, 1963), 187/2560.2.

(Xihu laoren) of the Southern Song (1127-1279) states that elephants were part of the daily morning court ritual.

> Foreign countries sent six elephants and two camels. Among them is a female elephant, which is called Little Girl Number Three. An elephant park was built outside the gate of Jian Bridge to settle them. Every day, they went with palace officials to the front of the palace to make a greeting, and returned after the court audience was finished. In front of them, there were teams of drum and gong players and thirty to forty people holding flags in multiple colours; on the back of each elephant was a person wearing a hat, holding a hook, and donning a purple gown. All the retinues wore robes and hats. On their way, they struck drums and gongs to guide the elephants back to the elephant park.[30]

It is doubtful if the elephants' procession was practised every day, as the morning court audience usually took place once every five days in the early hours of the morning.[31] Nonetheless, it is likely that elephant processions were regularly rehearsed and thus drew public attention. As an flaneur-like observer of urban life, the author, the so-called Elder from the West Lake, probably had ample opportunity to see these scenes.

Although reserved for court rituals, elephant watching became a source of entertainment, often taking place in imperial parks. In 1012, the court held a banquet at Yujing Garden, where exotic tributary animals including elephants and lions were on view.[32] In addition, these elephants were also considered a must-see especially for envoys from the northern countries. During the Northern Song, the viewing regularly took place at Yujin Garden. During the Southern Song, due to the limited space in the capital, the viewing took place at the camel stable, which also kept elephants. In 1145, when the camel stable was not available, elephants were brought to the hostel of the envoys.[33] The commoners would also have access to satisfy their curiosity. Serving staff tended the gardens; farmers cultivated crops and fruit trees. Moreover, city dwellers could visit imperial parks on special occasions or by paying the gatekeepers.[34] It was in this way that the Japanese monk Jōjin (1011–1081) was given a guided tour at the elephant stable in Ningling. Jōjin was shown seven elephants. Each elephant first bent its hind legs, then crouched and lowered its head, and finally made a trumpeting sound as instructed. After each step of this performance, the elephant keeper asked Jōjin to pay fifty coins, amounting to 350 coins for the whole tour. The performance of the elephants, as described by Jōjin, closely resembled their

30 Xihu laoren, *Fasheng lu*, in *Quansong biji* (Zhengzhou: Daxiang chubanshe, 2017), vol. 8.5, p. 324.

31 Zhao Sheng, *Chaoye leiyao*, 1b-2a.

32 Xu Song, *Song huiyao jigao*, Zhiguan 36.45.

33 Xu Song, *Song huiyao jigao*, Fangyu 3.17.

34 Wang Deli, "Songdai de Yujinyuan," *Huanghe keji daxue xuebao* 14 (2012.5): 76–80.

performance at the court ceremony. In 1073, an edict titled *Official Instructions for the Elephant Performance at the Southern Suburban Sacrifice* was issued summoning all seven elephants from Ningling to the capital to participate in the ritual.[35] The seven elephants Jōjin saw may have been the same ones that took part in the Southern Suburban Sacrifice. Significantly, the keeper's request for cash makes it clear that keeping elephants for the purpose of court ceremonies was turning into a commercial, monetised performance and a spectacle accessible to the wider public.

In the capital, elephants were undoubtedly a genuine urban spectacle. Elephants had been used in imperial parades as early as the Han dynasty. In the imperial inspection tour to the state shrines in Ganquan (in Shaanxi province) and Fenyin (in Shanxi province), an elephant-drawn carriage accompanied by thirteen musicians formed an impressive feature of the retinue of the emperor.[36] Elephants were used as draught animals, like horses. As mentioned above, seven elephants were summoned to the capital, but usually only six were used in the inspection tour; the seventh was probably a substitute.[37] Trained elephants usually formed the head of the procession (*yinjia*) in court-centred public activities, especially imperial processions.[38] In 968, six elephants were used in the Southern Suburban Sacrifice, and this number was maintained throughout the Northern Song dynasty.[39] According to *Song History*, in the imperial procession, each elephant was adorned with a wooden lotus seat on its back, a purple embroidered harness on its head, and brass bells and apricot leaf shaped plates on its chest and hips. It was also decorated with a crimson-coloured whisk made of yak's tail. Four soldiers guided the elephant, which carried a mahout from Nanyue (present-day Northern Vietnam) on its back. All soldiers wore patterned headwraps, tight red embroidered clothing, and silver belts.[40]

Although the *Illustration of the Imperial Guard of Honor* (Dajia lubu tu) does not include the elephants, the inscriptions accompanying this long hand scroll state that six elephants are placed at the head of the procession for events such as the Southern Suburban Sacrifice, followed by 5,481 officials and soldiers, seventy-one vehicles, 2,873 horses, two ponies, thirty-six oxen, 4,330 musical instruments, and ritual weaponry. Such an arrangement had practical reasons; with their massive weight and size they could test the span and supporting power of bridges. The procession for the Southern Suburban Sacrifice usually started on the winter solstice, considered to be the right

35 Xu Song, "Yangxiang suo," in *Song Huiyao jigao*, Zhiguan 23.3.

36 Liu Xin, *Xijing zaji* (Baojingtang congshu edition, Beijing: Zhili shuju, 1923), 2.11a–b.

37 Xu Song, "Yangxiang suo," in *Song Huiyao jigao*, Zhiguan 23.3.

38 Cheng Minsheng, "Songdai daxiang de ziran yu shehui shengtai": 1–14. Edward Schafer also discusses war elephants, see Schafer, *War Elephants in Ancient and Medieval China* (Leiden: Brill, 1957).

39 Ouyang Xiu comp., *Taichang yinge li*, in *Congshu jicheng chubian* (Shanghai: Shangwu yinshuguan, 1936), vol. 1044, 28/174.

40 Tuotuo, *Song shi*, 148/3461. Also Xu Song, "Yangxiang suo," in *Song Huiyao jigao*, Zhiguan 23.3, 23.9.

beginning of a new year. Mounted by mahouts from Nanyue, the elephants were made to draw large carts carrying musicians. They were placed right after the ritual guards and the palace guards. The ritual guards held the placard of the Minister of Rites and played compositions known as *The Ode to Inner Solemnity* (Zhongyan) and *The Ode to Outer Affairs* (Waiban); they were followed by the palace guards on horses in armour and rounds of regalia.[41] Taking place once every three years, the procession must have ignited the curiosity of the public. Even before this official 'performance,' the two months of rehearsals could have drawn more spectators.

> In years of great ritual, a rehearsal of carts and elephants started two months in advance. They marched from Xuande Gate to the outside of Nanxun Gate, and then returned. [...] There were seven elephants. In front of them dozens of vermilion flags and dozens of brass gongs and hand-drums were arrayed. The ritual started with two strikes of gongs, followed by three strikes of drums. Flag carriers wore purple garments and hats. Each elephant was shepherded by a mahout wearing a headwrap with crossed hanging bands and a purple gown. The mahout rode on the neck of the elephant, carrying a short-handled brass hook with a sharp blade. He hit the elephant when it disobeyed. When the elephants arrived in front of Xuande Gate, they turned around a few times to stand in line and were made to salute north. They were also capable of making greetings. Every day, the households related to the royal families and the households of noble origins called the elephants to their own residences to watch them, gifting them coloured silk. The Imperial Street was crowded with joyful visitors and observers. Little elephants made from rammed earth or wood powder as well as paper images of them were sold. People who had seen this brought them home as souvenirs.[42]

One gets the impression that the training of the elephants was more attractive than the actual performance. The elite and the commoners in the Song capital were equally enthusiastic about watching the rehearsals. The path from the Nanxun Gate to the Xuande Gate connected the inner palace and the liveliest marketplace in the capital. During the two months before the inspection tour, souvenirs turned the elephants into commodities and collectibles. The rehearsal of elephant processions continued in Lin'an (Hangzhou, in Zhejiang), the capital of the Southern Song, but only two elephants were used in the same ritual. They were displayed in front

41 Zhongguo guojia bowuguan ed., *Zhongguo guojia bowuguan guancang wenwu yanjiu congshu* (Shanghai: Shanghai guji chubanshe, 2007), p. 15. For an earlier illustration of the procession, see Patricia Ebrey, "Taking Out the Grand Carriage: Imperial Spectacle and the Visual Culture of Northern Song Kaifeng," *Asia Major* 12.1 (1999): 33–65.

42 Meng Yuanlao, *Dongjing meng Hua lu* (Beijing: Zhonghua shuju, 2006), 10/883. For the translation of the full passage, see Ebrey, "Taking Out the Grand Carriage," 38.

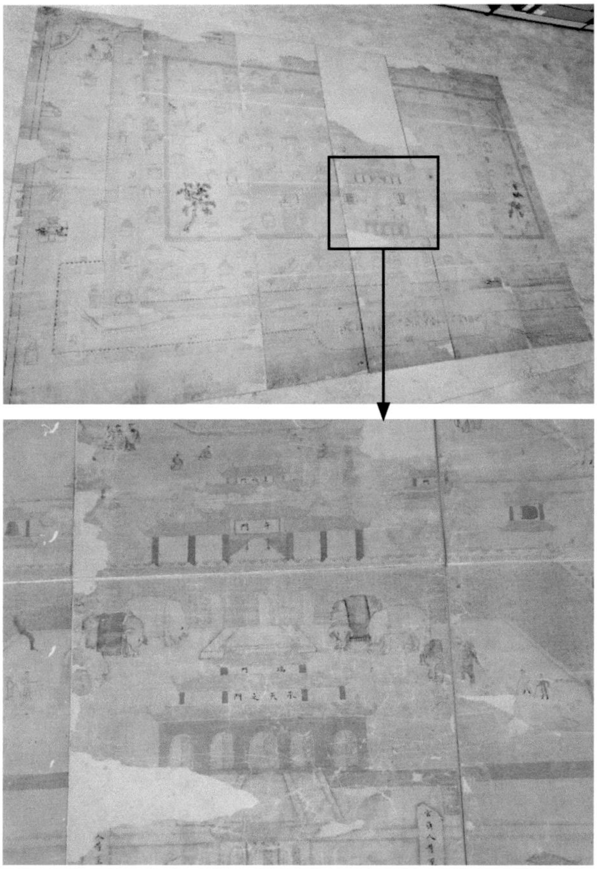

Figure 4.1: *Map of the Capital of the Ming Dynasty*. 217 cm x 160 cm. Colours on paper. Ming dynasty. The Courtesy of Leiden University Libraries. (a) detail of the elephant stable; (b) detail of Wu Gate.

of the Ancestral Temple and the Gate of Grandeur and Uprightness. Like in the Northern Song, spectators were more interested in the rehearsing of the elephants than in the solemn ceremony. Similarly, travellers to Lin'an (*waijing ren*) who had seen them returned to their homes outside Lin'an with figurines and paintings of elephants.[43] To use Susan Stewart's words, these souvenirs offered "the possibility of incorporation," in the sense that they moved the idea and image of elephants from the market space into the wider world and created an aura of urbanity.[44]

Grand processions continued during the Yuan and Ming dynasties. Kublai Khan (1260–1294) was known to travel annually in a carriage drawn by four elephants between two capitals, Xanadu (in Inner Mongolia) and Dadu.[45] The distance between the two capitals is about 450 kilometres. In a set of paintings depicting Emperor

43 Wu Zimu, *Mengliang lu* (Hangzhou: Zhejiang renmin chubanshe, 1981), p. 31.

44 Susan Stewart, *On Longing: Narratives of the Miniature, the Gigantic, the Souvenir, the Collection* (Durham, NC: Duke University Press, 1993), pp. 132–151.

45 Song Lian, *Yuan shi* (Beijing: Zhonghua shuju, 1976), 79/1974, 167/3925–3926.

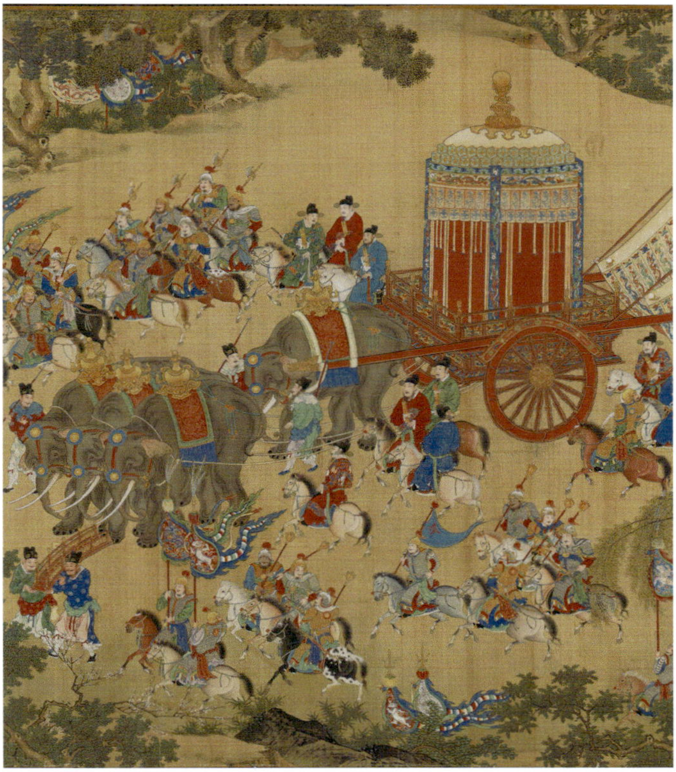

Figure 4.2: Anonymous, *Inspection Tour* (*Chujing tu*), detail. Colours on silk. 300 cm x 50 cm. Ming dynasty. Courtesy of The Palace Museum, Taipei.

Shenzong's (r. 1572-1620) inspection tour, we also see four elephants drawing a carriage, each of them carrying a vase on its back (Figure 4.2). As the word, 'vase' (*ping*) puns with 'peace,' and 'elephant' also means 'sign' or 'phenomenon,' the image of placing a vase on an elephant was understood as a 'sign of peace.'[46] Interestingly, the image features two attendants carrying a bridge on their shoulders, symbolising the elephants' role as the vanguard of the parade.

Elephants Bathing as an Urban Spectacle

As mentioned above, the elephant stable was built near Xuanwu Gate in the inner city close to the lively commercial neighbourhood in the southern part of Beijing (Figure 4.1a).[47] In the late Ming, when grand ceremonies were held at Fengtian Palace

46 Lin Wanxuan, "Zhizuo taiping – Qingdai gongting 'Taiping youxiang' chenshe," *Gugong wenwu yuekan* 457 (2021.4): 62–74.

47 Wu Hung, *Remaking Beijing: Tiananmen Square and the Creation of a Political Space* (Chicago, IL: University of Chicago Press, 2005).

on occasions such as the birthday of the emperor, New Year's Day, and the Winter Solstice, about thirty elephants were used: one carrying a treasure box, two drawing the jade carriage and two the grand carriage; four in front of Fengtian Gate, four in front of each of the side gates to the east and west, six at Wu Gate, four at Duan Gate, four at Chengtian Gate, and four on both sides of Chang'an Gate.[48] In quite a few surviving maps of the Ming capital, six elephants are the only type of animal shown (Figure 4.1b). At a regular morning court service, six elephants were used in front of Wu Gate.[49] This is depicted in a genre of figure painting typically entitled *Painting of Waiting for the Morning Court Hour* (Dailou tu). In the version housed in Nanjing Museum, six elephants are already standing in front of the closed Wu Gate; the official in the foreground is waiting for the gate to be opened (Figure 4.3). Images of six elephants standing between Duan and Wu Gates not only point to the morning court service, but also signify the authority of the palace.

Also in the late Ming, public interest had shifted from merely watching elephants to watching elephants bathing. Water is essential for elephants. They need to drink a large amount of water and regularly bathe in a river or lake.[50] Moreover, domesticated elephants, especially those used in rituals, had to be bathed and cleaned properly. During the Yuan dynasty, the elephants' first stable, as mentioned above, was close to the Grand Canal and their second stable in Qincheng had direct access to hot springs in winter.

Elephant bathing at Xuanwu Gate was a truly popular urban spectacle in the Ming dynasty. Written records often associate elephant bathing with the sixth day of the sixth month. That day had become significant for the Song court as the day on which Emperor Zhenzhong (r. 997–1022) was allegedly bequeathed a heavenly scripture in 1011.[51] However, although the court and Daoist temples held annual rituals to commemorate this day, it was not celebrated as a festival among commoners. During the Ming, the day was stripped of its religious meaning and acquired a larger impact on people's daily lives. It was turned into a special day on which both court and commoners exposed stored goods to the air or cleaned things with water. Elephant bathing was thus incorporated into the day's practices, which added to the festive character of the custom. According to an entry titled

48 Shen Shixing, *Da Ming huidian*, 228/3008.2. For more information about the use of elephants in Ming court rituals, see Liu Xiangxue, "Mingdai xunxiang wei kaolun," *Lishi yanjiu* (2011.1): 63. Only twenty-nine elephants are listed here; the one left was probably used as a substitute.

49 Occasionally, four elephants are represented. For more versions of these theme, see Huang Xiaofeng, "Zijingcheng de liming – Wanming de dijing jingguan yu guanliao xiaoxiang," in *Yu zaowu you: Wanming yishushi yanjiu (er)*, edited by Li Anyuan (Changsha: Hunan meishu chubanshe, 2017), pp. 7–52.

50 Thomas Trautmann, *Elephants and Kings: An Environmental History* (Chicago, IL: The University of Chicago Press, 2015), p. 26.

51 Tuotuo, *Song shi*, 8/147.

'The Sixth day of the Sixth Month' in *Collected Accounts Gathered in the Country during the Wanli Period* (Wanli yehuo bian),

> The sixth day of the sixth month was originally not a proper festival, but only a day when the Archive of the Imperial Household Department aired boxes of veritable records and imperially composed proses of all past emperors. Now it happens every year. As for the women who followed the custom of the time, if they washed their hair on this day, the hair never gets greasy or dirty. As for cats and dogs, they would also be bathed in the river. Elephants would also be bathed outside of the city on this day, and only once every year. They would copulate in the water, with the female facing up and the male facing down like human beings.[52]

Similar accounts may be found in several poems, prose texts, and notebooks (*biji*) of the late Ming dynasty.[53] However, the copulation mentioned above might not have happened at all, as domesticated elephants are almost never bred.[54] However, the author added a layer of fantasy to the sensation of the exoticness of the scene. In the late Ming notebook of urban life in Beijing, *An Overview of Sights and Affairs of the Imperial Capital* (Dijing jingwu lue) added that chariots and clothing were aired in the palace, and books and cotton garments of commoners were also put under the sun. It also documented how elephants were bathed:

> With flags and drums, the Embroidered-Uniform Guard Unit led the elephants out of Shuncheng Gate (aka Xuanwu Gate). After the bathing sluice had been opened, the elephants entered the river one by one, like the collapse of a black mountain. After a while, they returned with their heads raised. Their trunks sucked and sprayed water, powerful like a dragon. The elephant mahouts held the reins and sat on their backs; one could see their bound-up hair from time to time. On each bank, tens of thousands of people were watching.[55]

The above two quotations from *Collected Accounts Gathered in the Country during the Wanli Period* and *An Overview of Sights and Affairs of the Imperial Capital* offer us a glimpse to understand how the day gained a festive dimension. Since the sixth day of the sixth month was considered the hottest day of the year, sun and water on that day were believed to be particularly efficient for cleaning and washing.

52 Shen Defu, *Wanli yehuo bian*, pp. 619-620.

53 Liu Dong and Yu Yizheng, *Dijing jingwu lue* (Beijing: Beijng guji chubanshe, 1983), p. 78. Xie Jie and Shen Yingwen, *Wanli Shuntian fu zhi* (1593 edition), 1/15b.

54 Thomas Thautmann, *Elephants and Kings: An Environmental History* (Chicago, IL: The University of Chicago Press, 2015), p. 60.

55 Liu Dong and Yu Yizheng, *Dijing jingwu lue*, p. 69.

Since the city moats where the elephant bathing took place were connected to Lake Taiye to the north, water was easily accessible. The Xuanwu Gate was located in the south of the inner city, the concentrated neighbourhood of commoners. In addition to elephants, horses from the imperial stables were also bathed in rivers and lakes.[56] However, it was the bathing of elephants that turned the custom into a spectacle. As mentioned above, the elephants must have been regularly cleaned for court processions, but their bathing on this special day was a different issue. It was intended to be a secular, non-political event. Initiated by the Elephant Training Guard, a unit of Embroidered-Uniform Guard, it only became an urban event because of the vast public interest. In the city moats, the elephants ceased to be the embodiment of political rituals but were brought into a vernacular, urban context. In comparison with elephants in court rituals that were often associated with the palace gates and imperial processions, elephants that were bathed in the city moats and watched by enthusiastic urbanites were clearly the focus of an urban spectacle.

Elephant Bathing in Visual Images

The theme of elephant bathing in art historical narratives first appears to have connections with India. Unlike in medieval China, elephants in ancient India were trained to serve the court and played a significant role in economic and religious life.[57] The *Xuanhe Catalogue of Paintings* (Xuanhe huapu), completed in 1120 under the auspices of Emperor Huizong (r. 1100–1126), lists four paintings titled *Scrubbing an Elephant* (Saoxiang tu) by Zhang Sengyou (479–?), Yan Liben (601–673), Sun Wei (active in the late ninth century), and Sun Zhiwei (tenth to eleventh century).[58] None of these paintings have survived, but in the late Southern Song, Zhou Mi (1232–1308), a renowned scholar and connoisseur, had the opportunity to see the painting attributed to Yan Liben. According to Zhou, Yan drew a foreign monk wetting a broom to wash a black elephant and an attendant standing next to him scrubbing the elephant with a cloth. Although in the *Xuanhe Catalogue*, all four paintings are listed in the category of 'Daoism and Buddhism', Zhou Mi admitted that he had no knowledge of the original meaning of the painting. The earliest visual example of this theme is a rubbing of a Song inkstone collected by the Qing court (Figure 4.4). The inkstone bears the mark of the Xuanhe reign era. Unlike Yan Liben's rendition of the scene, the elephant here is only accompanied by a foreign-looking attendant. The Qing description of the inkstone identifies him as a mahout (*xiangnu*) but based

56 Yu Minzhong, *Rixia jiuwen kao*, in *Yingyin Wenyuange Siku quanshu* (Taipei: Taiwan shangwu yinshuguan, 1983), vol. 497, 54/39.1 (765).

57 Chen Huaiyu, *In the Land of Tigers and Snakes: Living with Animals in Medieval Chinese Religions* (New York: Columbia University Press, 2023), pp. 23, 29.

58 Zhao Ji, Yu Jianhua ed., *Xuanhe huapu* (Beijing: Renmin chubanshe, 1964).

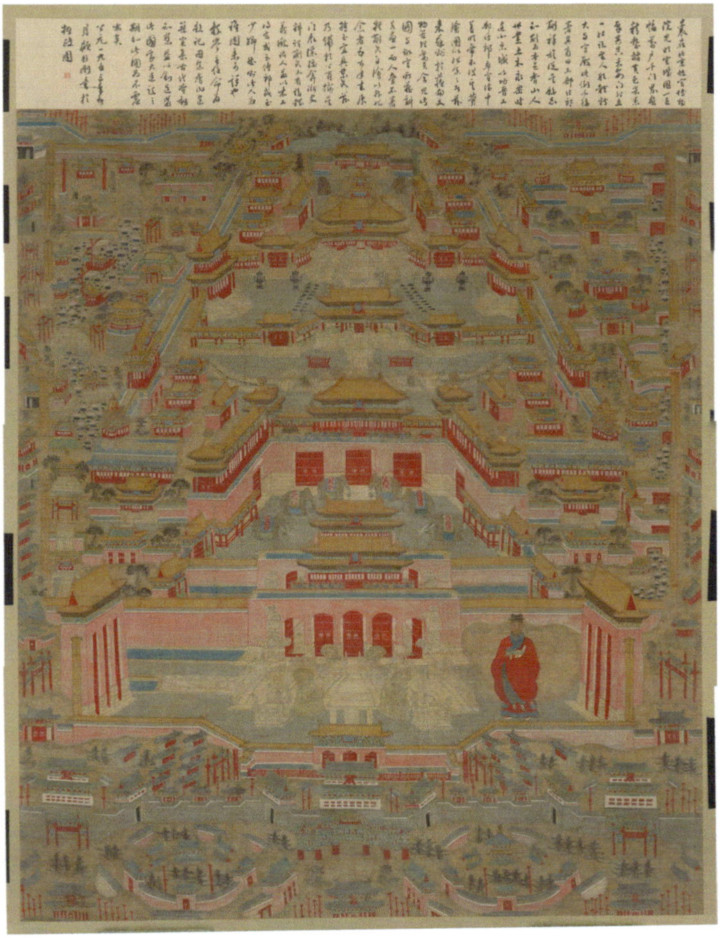

Figure 4.3: Anonymous, *Painting of Waiting for the Morning Court Hour* (Dailou tu); also titled *Painting of the Inner Palace* (Beijing gongcheng tu). Colours on silk. 100 cm x 189 cm. Ming dynasty. Courtesy of Nanjing Museum, Nanjing.

on the brass headband the figure is wearing, he could also be identified as a monk. The pictorial design on the rubbing has a dynamic composition: the elephant turns its head towards the attendant who is scrubbing it with a piece of cloth. The cloud pattern in the upper register and the two lotus flowers in the lower register point to the auspicious nature of the animal. However, just like Zhou Mi, we cannot deduce the metaphorical meanings of the theme, if there are any, from the rubbing or the title of the paintings in the *Xuanhe Catalogue*. It is, though, safe to conclude the following: first, in addition to an elephant, paintings of this kind always include a foreign-looking attendant or a foreign-looking monk, or sometimes both; second, the elephant is a rather mundane animal, which is black and with two tusks.

Paintings entitled *Scrubbing an Elephant* and *Elephant Bathing* (*xixiang*) gained popularity in the late Ming. Most of these paintings were attributed to a few renowned painters, including Qian Xuan (1239–1299), Ding Yunpeng (1547–1628), and Cui Zizhong (1597–1644). Art historians agree that most paintings attributed

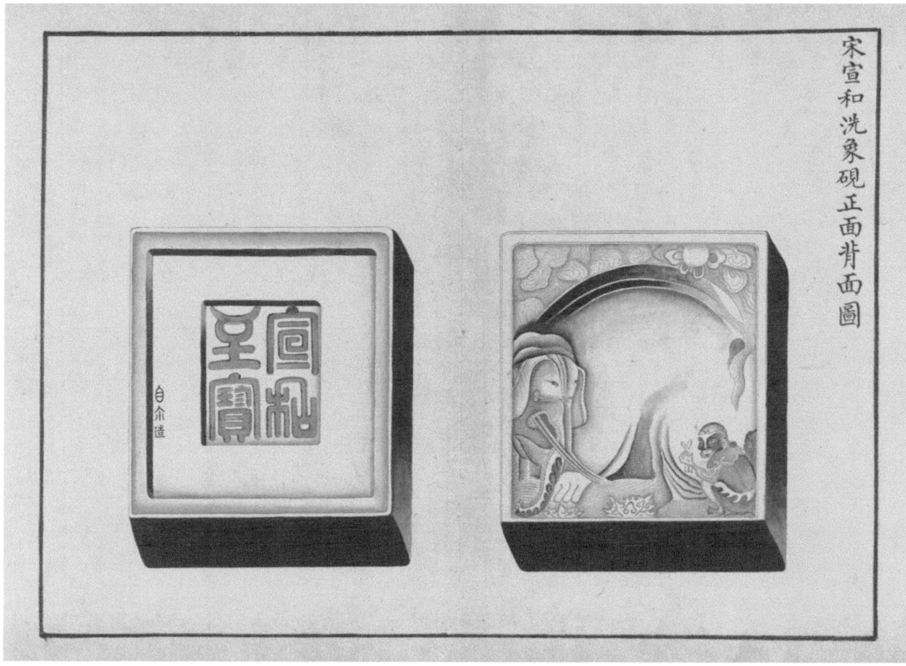

Figure 4.4 Anonymous, Drawings of the "Elephant Bathing Inkstone of the Xuanhe Era in the Song Dynasty" (Song Xuanhe Xixiang yan), front and back, in Yu Minzhong et al., *Imperially Endorsed Catalogue of Inkstones from the Chamber of Western Purity* (Qinding Xiqing yanpu), *juan* 7, p. 35. Courtesy of the Institute of History and Philology, Academia Sinica.

to Qian Xuan were counterfeited in the late Ming (Figure 4.5).[59] Paintings attributed to Qian shared the same composition: a monk, a white elephant, and a few mahouts; paintings attributed to Ding Yunpeng and Cui Zizhong also include a bodhisattva. Although the theme of elephant bathing became popular in the late Ming, its meaning was never articulated in a concrete, definitive manner. Until the early Qing, the identity of the bodhisattva remained unclear, as it was sometimes identified as Mañjuśrī (Wenshu) and other times Samantabhadra (Puxian).[60] It

59 Julia Andrews, "The Significance of Style and Subject Matter in the Painting of Cui Zizhong" (Ph.D. Dissertation, University of California, Berkeley, 1984); Lan Yujing, "Wanming Qingchu de *Caoxiang tu* yanjiu" (MA Thesis, Taiwan Normal University, Taipei, 2009). More recent research include Wang Zhongxu, "Chuan Qian Xuan Xiangxiang tu zhenwei kaobian – Jianji wang Ming zhiqian dui Xixiang tu de renshi," *Gugong bowuyuan yuankan* (2017.3): 107–124; Zhu Wenqi, "Ming Wanli Xixiang tu yanjiu—Yi Ding Yunpeng he Cui Zizhong weili" (MA Thesis, Nanjing Art Academy, 2020), pp. 1–3.

60 Wang Zhongxu, "Qianlong yurong hua jiqi kongxiang guan – Ding Guanpeng Saoxiang tu, Lang Shining Qianlong guanhua tu," *Gugong xuekan* 18(2017): 97–108. Although the meaning of elephants was more diverse in Buddhist texts, their visual representations in paintings as well as on sculpture are mainly associated with two themes: (1) the birth of Shakyamuni, as Queen Maya, the future mother of Shakyamuni, had a dream in which a white elephant with six tusks entered her womb before the birth

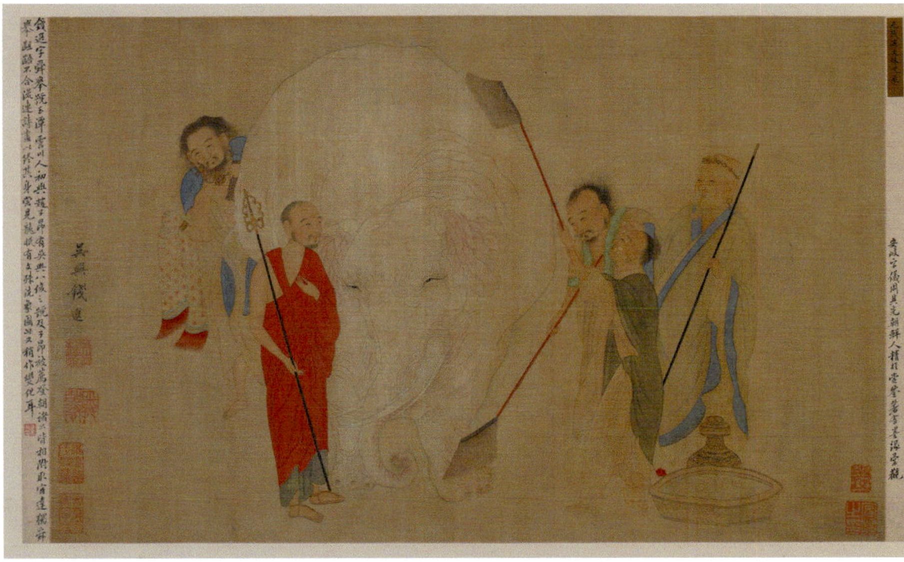

Figure 4.5: Attributed to Qian Xuan, *Elephant Bathing* (*Xixiang tu*). Colours on paper. 124 x 52 cm. Replica made during the Ming dynasty. Courtesy of the Freer Gallery of Art, National Museum of Asian Art. Washington D.C.

was also in the Qing that *xixiang* or *saoxiang* came to be understood as a Buddhist metaphor for the purification of the mind, as the character 'elephant' (*xiang* 象) and 'phenomenon/form' were often used interchangeably.[61]

Nevertheless, the underlying connections between the practice and the visual representation of elephant bathing are worth noting. The paintings gained popularity when elephant bathing came to be a popular scene in the capital. The Japanese woodblock printed book, *Illustrations of Famous Places in China* (Morokoshi meisho zue), lists elephant bathing as one of the most eye-catching scenes in the capital. It provides three images, including elephants in processions, elephant bathing, and tools used by mahouts (Figure 4.6). Although the scene of elephant bathing is not set outside the city wall, it does catch the liveliness of the spectacle. The image of the tools further augments the sense of reality. In the Ming–Qing contexts, even

of the buddha; (2) the bodhisattva Samanthabhadra (*Puxian pusa*), as the six-tusk white elephant is believed to be the attribute animal of Samanthabhadra. Kumārajīva, *Miaofa lianhua jing*, in *Taishō shinshū Daizōkyō*, edited by Junjirō Takakusu and Kaigyoku Watanabe (Tokyo: Taishō Shinshū Daīzōkyō Kankōkai, 1960–1978) T9.262, 28.61a–b. The white colour of an elephant was not only associated with Buddhism, but was also understood auspicious in general. See Zou Zhenhuan, "Dongya shijie de 'xiang' ji," in *Zaijian yishou: Mingqing dongwu wenhua yu zhongwai jiaoliu* (Shanghai: Shanghai guji chubanshe, 2021), pp. 153–154.

61 Wang Zhongxu, "Chuan Qian Xuan Xiangxiang tu zhenwei kaobian – Jianji wang Ming zhiqian dui Xixiang tu de renshi," pp. 123–124.

if the elephant is depicted as a sacred being, with a white skin or six tusks, it is always attended by mahouts and the image includes a river. In fact, both Ding Guanpeng and Cui Zizhong must have seen the bathing of elephants, as Ding served as a professional painter at the imperial court in Beijing and Cui Zizhong lived in Beijing for decades during the Chongzhen Era (1628–1644). A poem that the Ming loyalist Wu Weiye (1609–1671) inscribed on a painting entitled *Elephant Bathing* by Cui Zizhong might be enlightening. Wu composed this poem in 1653, nine years after the collapse of the Ming dynasty, lamenting the good times of enjoying Cui's paintings and the spectacular bathing scene.

> Throughout his life, [Cui] was skilled in painting the bathing of elephants,
> When in the mood, he wielded his brush over the unfolded scroll.
> The red cover on the elephant has the decorative patterns from Erhai,[62]
> With its white tusks, it stands in front of Hanyuan Hall.[63]
> To Chengtian Palace Gate, it drove the emperor's cart,
> With a banner bearing the *luan* bird, the sun, and the moon.
> Beneath the two pavilions, roosters crowing, bells striking[64]
> The elephant stood firm, towering like Mont Kunlun.
> Shouldering aside the stablemen and imperial guards,
> The commoner Cui held paper and brush in his hands.
> He looked up, seeing the trained elephants coming down the imperial street,
> After returning home, he contemplated for ten days.
> ...Following the customs of the capital, people came to see the bathing of elephants,
> The jade river in the spring was pure and clean.
> Black mahouts were barefoot, each fastening two brooms on his back,
> All the streets were filled with carts of gentlemen and ladies.
> The elephants lifted their trunks, making thunder-like sounds in the northern pavilion;
> Water was trampled, being splashed like snow in the West Mountain.
> The painting that Cui completed was hung in the market of the capital,
> Viewers by the roadside marvelled at its excellence.
> Being unique, Cui did not like making paintings to meet the demand;
> The price was raised high, and only the powerful and the rich could snatch them...[65]

62 Erhai is located in Kunming (Yunnan province), one of the places where elephants were native.

63 Hanyuan Hall was the main hall in the Tang imperial palace. Here it refers to the imperial palace in general.

64 The 'two pavilions' usually refers to the pavilions in front of the imperial palace, and thus is the synecdoche for imperial palace.

65 Wu Weiye, *Wu Meicun quanji* (Shanghai: Shanghai guji chubanshe, 1990), vol. 1, 11/305–307.

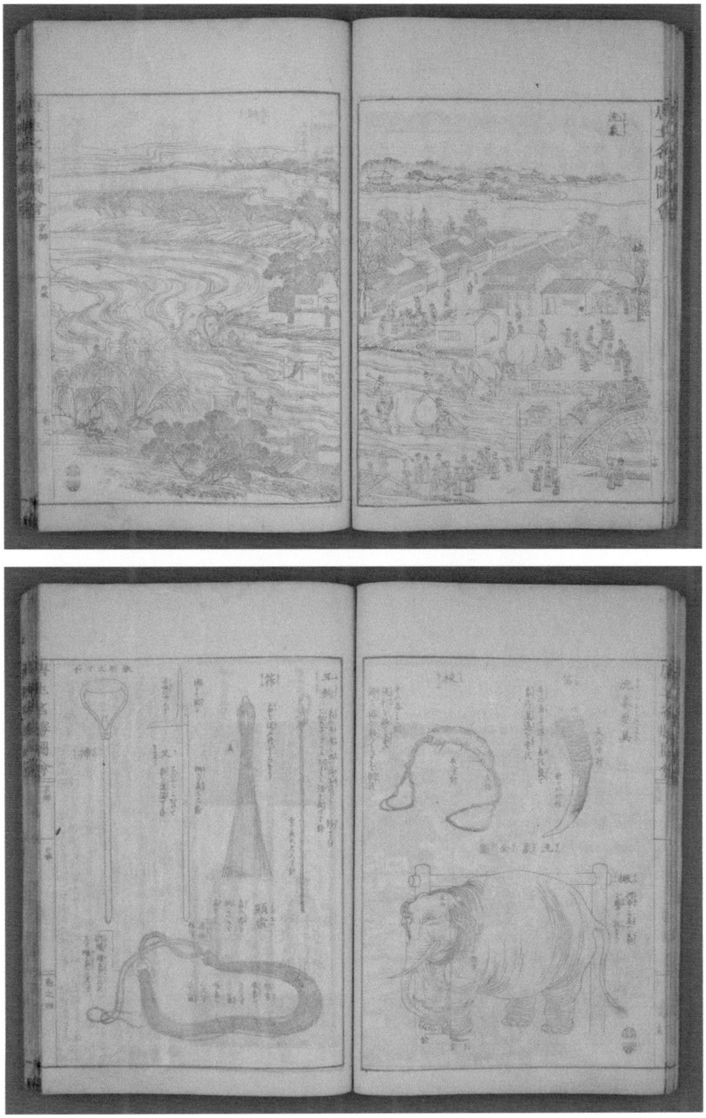

Figure 4.6: Okada Gyokuzan, Oka Yugaku, and Ohara Toya, *Illustrations of Famous Places in China* (*Morokoshi meisho zue*) (Kawachiya Kichibei, 1806), vol. 4, 14a-b, 15a-b.

The poem can only be found in Wu Weiye's anthology. It is therefore unclear which painting this poem refers to. However, all extant paintings attributed to Cui have the same transcendental theme, which includes a six-tusk elephant bathed by two mahouts, a bodhisattva, and a monk (Figure 4.7). It is very likely that Wu Weiye's poem was inscribed on a similar painting. Wu's poem clearly indicates that Cui's painting was inspired by the actual scene of elephant bathing in the capital. Cui turned the ordinary black elephant into a white, sacred one, and has the barefoot, black mahouts carrying brooms. Just like the souvenirs during the Song, Cui's paintings were sold on the market.

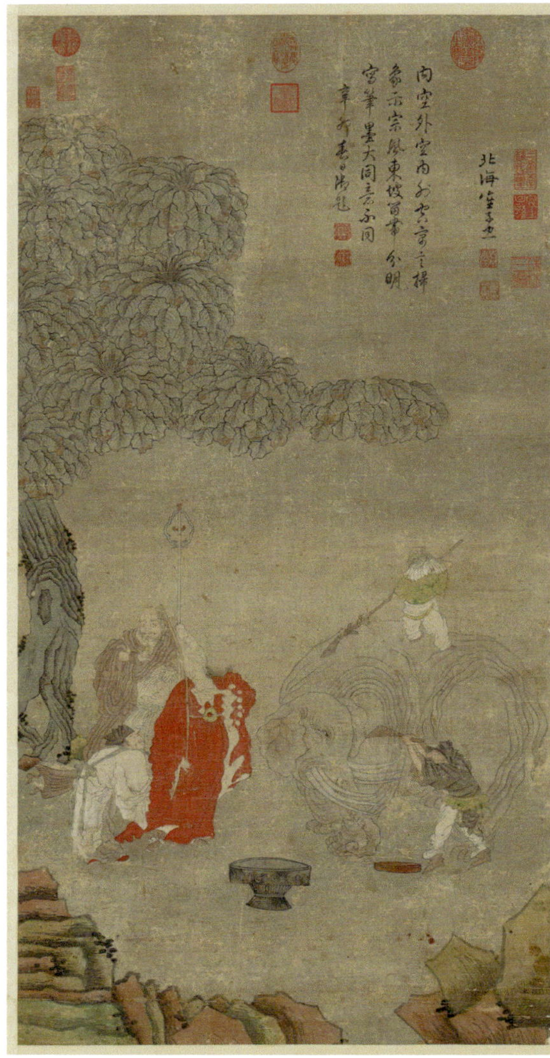

Figure 4.7: Cui Zizhong, *Scrubbing the Elephant* (*Saoxiang tu*). Colours on silk. 50 x 152 cm. Ming dynasty. Courtesy of the Freer Gallery of Art, National Museum of Asian Art. Washington D.C.

Conclusion

After wild elephants had moved to the southern edge of the Song empire, they were 'returned' as tamed, tributary animals to the capitals Kaifeng, Hangzhou, and Beijing. Their performance in court rituals, including imperial processions on various occasions and at the morning courts, was meant to be a visual confirmation and reinforcement of imperial power. It was in this context that tamed elephants were sent into the capital, and policies and facilities were developed to accommodate them. As a result, elephants entered public consciousness as a political phenomenon. However, the display of elephants could only achieve such an effect when it was seen by the general public. Among the urbanites in the capital, the watching of elephants increasingly broke away from the political agenda of the court. Already in the Song, people were drawn to the imperial processions, but they were also keen on seeing the elephants in training or rehearsal and enthusiastically bought elephant souvenirs. The decoupling between the political meaning of elephants and their popular image became more evident in the Ming when elephant bathing rather than their service in the imperial context ignited the interest of the public. The performative scene of their bathing contributed to turning the sixth day of the sixth month into a festival. It was originally the traditional cleaning and airing day, but elephant bathing turned it into a carnival-like festival. Paintings of elephant bathing could also be understood along this line. The motif, first appearing in the Song, includes a foreign-looking mahout or monk beside a river. From the late Ming onwards, the motif combined bodhisattvas and was assigned religious meanings, but, as Wu Weiye's poem shows, it still had an underlying relationship with the popular practice of elephant bathing.

The political, religious, or cultural meanings of elephants could only make sense when they or their images were watched by spectators. In the eyes of the city dwellers, the exotic animal had many faces. They were sacred animals in Buddhist anecdotes and scriptures, political animals that were brought from afar to serve the court, and exotic animals to be watched for fun. In comparison with the elephants guarding the palace and leading the emperor's processions, the elephants in the city moats in apolitical, urban contexts were more ready to show their true nature.

Glossary

An	安
Beijing gongcheng tu	北京宮城圖
Bian	汴
biji	筆記
Chang'an	長安
Chengtian	承天
Chongxin	崇新
Cui Zizhong (1597–1644)	崔子忠
Chujing tu	出警圖
Dadu	大都
Dailou tu	待漏圖
Dajia lubu tu	大駕鹵簿圖
Dijing jingwu lue	帝京景物略
Duan	端
Erhai	洱海
Fansheng lu	繁勝錄

Fengtian	奉天
Fenyin	汾陰
Fu	復
Ganquan	甘泉
Guangdong	廣東
Guangzhou	廣州
Haizi	海子
Hanyuan	含元
Hengzhou	橫州
Huizong (r. 1100–1126)	徽宗
Jian	薦
Jiangdu	江都
Jinshui	金水
Jinyi wei	錦衣衛
Jishuitan	積水潭
Jōjin (1011–1081)	成尋
Huangpi	黃陂
Jianlong	建隆
Jiaozhi	交趾
Lin’an	臨安
luan	鸞
Lingnan	嶺南
Luchuan tusi	麓川土司
Meng Yuanlao	孟元老
Ningling	寧陵
Nanxun (Gate)	南薰門
Puxian pusa	普賢菩薩
Qian Xuan (1239–1299)	錢選
Qincheng	秦城
qing	頃
Saoxiang tu	掃象圖
Shaozhou	韶州
She suo	射所
Shenzong (r, 1572-1620)	神宗
Shuncheng	順承
Siming	思明
Song shi	宋史
Sun Wei (active late ninth century)	孫微
Sun Zhiwei (tenth–eleventh century)	孫之微
Taiping	太平
Taiye	太液

Tang	唐
Wanli yehuo bian	萬曆野獲編
Wen Huanran	文煥然
wei	衛
Wu Weiye (1609–1671)	吳偉業
Wu Yue	吳越
Wuxing zhi	五行志
Waiban	外辦
xiangfang	象房
xiangnu	象奴
xixiang	洗象
Xiang	襄
Xihu laoren	西湖老人
Xiong Mengxiang (fourteenth century)	熊夢祥
Xuanhe huapu	宣和畫譜
Xuanwu	宣武
Xunxiang suo	馴象所
Xunxiang wei	馴象衛
Yan Liben (601–673)	閻立本
Yangxiang suo	養象所
Yinjia	引駕
Yujin	玉津
Zhenzhong (r. 997–1022)	真宗
zhang	丈
Zhang Sengyao (479–?)	張僧繇
Zhongyan	中嚴
Zhou Mi (1232–1308)	周密
zhihui shi	指揮使
ziran (Chin.)/*shizan* (Jap.)	自然
zi zhi jingshi	自至京師

Bibliography

Primary Sources

Kumārajīva 鳩摩羅什, trans. *Miaofa lianhua jing* 妙法蓮華經. In Takakusu Junjirō 高楠順次郎 and Watanabe Kaigyoku 渡部海旭 eds., *Taishō shinshū Daizōkyō* 大正新修大藏經. Tokyo: Taishō Shinshū Daīzōkyō Kankōkai, 1960–1978.

Liu Dong 劉侗 and Yu Yizheng 于奕正. *Dijing jingwu lue* 帝京景物略. Beijing: Beijing guji chubanshe, 1983.

Li Longji 李隆基 comp. *Tang liu tian* 唐六典. In *Yingyin Wenyuange Siku quanshu*, vol. 595. Taipei: Taiwan shangwu yinshuguan, 1983.
Li Tao 李燾. *Xu zizhi tongjian changbian* 續資治通鑑長編. Beijing: Zhonghua shuju, 2004.
Liu Xin 劉歆. *Xijing zaji* 西京雜記. Baojingtang congshu 抱經堂叢書 edition. Beijing: Zhili shuju, 1923.
Liu Xu 劉昫. *Jiu Tang shu* 舊唐書. Beijing: Zhonghua shuju, 1975
Meng Yuanlao 孟元老. Yi Yongwen 伊永文annotated. *Dongjing meng Hua lu jianzhu* 東京夢華錄箋注. Beijing: Zhonghua shuju, 2006.
Ming shilu 明實錄. Taipei: Zhongyang yanjiuyuan, 1963.
Okada Gyokuzan 岡田玉山, Oka Yugaku 岡熊嶽, and Ohara Toya 大原東野. *Morokoshi meisho zue* 唐土名勝図會. Kawachiya Kichibei, 1806.
Ouyang Xiu 歐陽修. *Taichang yinge li* 太常因革禮. In Congshu jicheng chubian 叢書集成初編, vol. 1043-1046. Shanghai: Shangwu yinshuguan, 1936.
Shen Defu 沈德符. *Wanli yehuo bian* 萬曆野獲編. Beijing: Zhonghua shuju, 1997.
Shen Shixing 申時行. *Da Ming huidian* 大明會典. Taipei: Guofeng chubanshe, 1963.
Song Lian 宋濂. *Yuan shi* 元史. Beijing: Zhonghua shuju, 1976.
Tuotuo 脫脫. *Song shi* 宋史. Beijing: Zhonghua shuju, 1977.
Wang Qinruo 王欽若. *Cefu yuangui* 冊府元龜. Beijing: Zhonghuashuju, 1994.
Wu Weiye 吳偉業. *Wu Meicun quanji* 吳梅村全集. Shanghai: Shanghai guji chubanshe, 1990.
Wu Zimu 吳自牧. *Mengliang lu* 夢粱錄. Hangzhou: Zhejiang renmin chubanshe, 1981.
Xie Jie 謝傑 and Shen Yingwen 沈應文. *Wanli Shuntian fu zhi* 萬曆順天府志. 1593 edition.
Xihu laoren 西湖老人. *Fansheng lu* 繁勝錄. In *Quansong biji* 全宋筆記. Zhengzhou: Daxiang chubanshe, 2017.
Xiong Mengxiang 熊夢祥. *Xi jin zhi jiyi* 析津志輯佚. Beijing: Beijing guji chubanshe, 1983.
Xu Song 徐松. Wang Deyi 王德毅 ed. *Song Huiyao jigao* 宋會要輯稿. Taipei: Academia Sinica, Chengdu: Sichuan daxue, Cambridge, MA: Harvard University, 2008. (https://hanchi.ihp.sinica.edu.tw/ihp/hanji.htm)
Yu Minzhong 于敏中. *Rixia jiuwen kao* 日下舊聞考. In *Yingyin Wenyuange Siku quanshu,* vols. 497-499. Taipei: Taiwan shangwu yinshuguan, 1983.
Yu Minzhong 于敏中 et al. *Qinding Xiqing yanpu* 欽定西清硯譜, in *Yingyin Wenyuange Sikuquanshu*, vol. 843.
Zhang Tingyu 張廷玉. *Ming shi* 明史. Beijing: Zhonghua shuju, 1974.
Zhao Ji 趙佶. Yu Jianhua 俞劍華 ed. *Xuanhe huapu* 宣和畫譜. Beijing: Renmin chubanshe, 1964.
Zhao Sheng 趙昇. *Chaoye leiyao* 朝野類要. Zhi bushuzhai edition, 1813.

Secondary Sources

Andrews, Julia. "*The Significance of Style and Subject Matter in the Painting of Cui Zizhong.*" Ph.D. dissertation, University of California, Berkeley, 1984.
Chen, Huaiyu. *In the Land of Tigers and Snakes: Living with Animals in Medieval Chinese Religions*. New York: Columbia University Press, 2023.

Cheng Minsheng 程民生. "Songdai daxiang de ziran yu shehui shengtai 宋代大象的自然與社會生態," *Zhongyuan wenhua yanjiu* 中原文化研究 (2021:3): 48–57.

Ebrey, Patricia. "Taking Out the Grand Carriage: Imperial Spectacle and the Visual Culture of Northern Song Kaifeng," *Asia Major* 12.1 (1999): 33–65.

Elvin, Mark. *The Retreat of the Elephants: An Environmental History of China*. New Haven, CT: Yale University Press, 2004.

Huang Xiaofeng 黄小峰. "Zijingcheng de liming – Wanming de dijing jingguan yu guanliao xiaoxiang 紫禁城的黎明—晚明的帝京景觀與官僚肖像, in Li Anyuan 李安源 ed., *Yu zaowu you: Wanming yishushi yanjiu (er)* 與造物游：晚明藝術史研究 (二). Changsha: Hunan meishu chubanshe, 2017, pp. 7–52.

Huang Yong 黃永. "Mingdai jingshi chaoxiang laiyuan kao 明代京師朝象來源考," *Gugong xuekan* 故宮學刊 (2008): 159–171.

Lan Yujing 藍御菁. "Wanming Qingchu de *Caoxiang tu* yanjiu 晚明清初的《掃象圖》研究." MA Thesis, Taiwan Normal University, Taipei, 2009.

Lin Wanxuan 林宛萱. "Zhizuo taiping – Qingdai gongting 'Taiping youxiang'chenshe 製作太平—清代宮廷「太平有象」陳設," *Gugong wenwu yuekan* 故宮文物月刊 457 (2021.4): 62–74.

Liu Xiangxue 劉祥學. "Mingdai xunxiang wei kaolun 明代馴象衛考論," *Lishi yanjiu* 歷史研究 (2011.1): 51–66.

Ohnuma, Reiko. *Unfortunate Destiny*. New York: Oxford University Press, 2017.

Schafer, Edward. *War Elephants in Ancient and Medieval China*. Leiden: Brill, 1957.

Stewart, Susan. *On Longing: Narratives of the Miniature, the Gigantic, the Souvenir, the Collection*. Durham, NC: Duke University Press, 1993.

Szonyi, Michael. *The Art of Being Governed: Everyday Politics in Late Imperial China*. Princeton, NJ: Princeton University Press, 2017.

Trautmann, Thomas. *Elephants and Kings: An Environmental History*. Chicago, IL: The University of Chicago Press, 2015.

Wang Deli 王德利. "Songdai de Yujinyuan 宋代的玉津院," *Huanghe keji daxue xuebao* 黃河科技大學學報 14 (2012.5): 76–80.

Wang Zhongxu 王中旭. "Chuan Qian Xuan Xiangxiang tu zhenwei kaobian – Jianji wang Ming zhiqian dui Xixiang tu de renshi 傳錢選《洗象圖》真僞考辨 – 兼及晚明之前對《洗象圖》的認識," *Gugong bowuyuan yuankan* 故宮博物院院刊 (2017.3): 107–124.

Wang Zhongxu 王中旭. "Qianlong yurong hua jiqi kongxiang guan – Ding Guanpeng Saoxiang tu, Lang Shining Qianlong guanhua tu 乾隆御容畫及其空相觀—丁觀鵬《掃象圖》、郎世寧《乾隆觀畫圖》研究," *Gugong xuekan* 故宮學刊 18(2017): 97–108.

Wen Huanran 文煥然 et al. *Zhongguo lishi shiqi zhiwu yu dongwu bianqian yanjiu* 中國歷史時期植物與動物變遷研究. Chongqing: Chongqing chubanshe, 1995.

Wu, Hung. *Remaking Beijing: Tiananmen Square and the Creation of a Political Space*. Chicago, IL: University of Chicago Press, 2005.

Yu, Hui-Chun. "When There Is Peace, There Are Elephants." In Marina Siebert, Chen Kai Jun, and Dorothy Ko eds., *Making the Palace Machine Work: Mobilizing People, Objects, and Nature in the Qing Empire*. Amsterdam: Amsterdam University Press, 2021, pp. 271–290.

Zhongguo guojia bowuguan 中國國家博物館 ed. *Zhongguo guojia bowuguan guancang wenwu yanjiu congshu* 中國國家博物館館藏文物研究叢書. Shanghai: Shanghai guji chubanshe, 2007.

Zhu Kezhen 竺可楨. *Tiandao yu renwen* 天道與人文. Beijing: Beijing chubanshe, 2006.

Zhu Wenqi 朱雯琦. "Ming Wanli Xixiang tu yanjiu – Yi Ding Yunpeng he Cui Zizhong weili 明代萬曆洗象圖研究—以丁雲鵬和崔子忠爲例." MA Thesis, Nanjing Art Academy, 2020.

Zou Zhenhuan 鄒振環. *Zaijian yishou: Mingqing dongwu wenhua yu zhongwai jiaoliu* 再見異獸：明清動物文化與中外交流. Shanghai: Shanghai guji chubanshe, 2021.

About the Author

Fan Lin is an art historian at the Institute of Area Studies at Leiden University. She obtained a PhD in Art History and East Asian Studies from McGill University, Montreal. Her research interests focus on mapmaking and material culture in Middle-Period China. Her recent publications include "The Shadow of Prosperity. Fake Goods and Anxiety in Song Urban Space" (2019), "Knowledge, Power, and Technology. Diagrams of Troop Formation in Early Song Military History" (2021), and "Bifurcated Memory. A Cultural Biography of the Porcelain Pagoda of Nanjing" (2022).

5. A Pair of Camels in Edo Japan: Representation and Discourse

Hiroyuki Suzuki

Abstract: This chapter examines the striking reception of a pair of single-humped camels in Edo Japan. The Dutch East India Company brought the camels to Nagasaki in 1821 as diplomatic gifts to the Tokugawa shogun. Rejected by the shogun, the camels were displayed in public shows around the country for years, attracting enormous attention. The camel shows not only provided a wide audience with information, both genuine and fabricated, but also inspired scholars, intellectuals, writers, and painters to produce novels, verses, songs, essays, and paintings. The camels' reception in Edo society generated a wealth of connotations concerning camels, which also reflected people's general view of animals.

Keywords: camel, diplomatic gift, *misemono* show, Edo culture, Maruyama Ōshin (1790–1838)

Introduction[1]

Focusing on a pair of camels that a Dutch ship brought to Nagasaki in 1821, this chapter examines the remarkable ways in which the animals were received by the urban dwellers of Edo Japan. In the sixth month of that year, a pair of dromedaries or single-humped camels (*Camelus dromedarius*) arrived on Deshima Island. *A Study of Public Shows* (Misemono kenkyū, 1928) by Asakura Musei (1877–1927) suggests that the *opperhoofd* or chief merchant of the Dutch East India Company (hereafter VOC) at the trading post in Nagasaki, Jan Cock Blomhoff (1779–1853), sold

1 This essay was based on a paper presented at the international symposium "Shifting Perspectives on Media and Materials in Early Modern Japan" at Djam Lecture Theatre on 4–5 July 2015, at SOAS, University of London. For this paper, the author revised his article "Rakuda o egaku: Maruyama Ōshin hitsu rakuda-zu o megutte" (To Depict the Camel: Problems Involved in "Camels" by Maruyama Ōshin), *Bijutsu kenkyū* 338 (1987), pp. 16–34.

Fan Lin and Doreen Mueller (eds), *Picturing Animals and Plants in Early Modern China and Japan. Innovation, Experiments, and Anxieties*. Taylor & Francis Group, 2025
DOI: 10.5117/9789048559091_CH05

the camels to an impresario[2] who orchestrated a series of shows exhibiting these strange animals in major cities such as Osaka, Kyoto, Edo, and Nagoya. Everywhere, a great number of local citizens came to enjoy the show. At the same time, the sight of the camels confused intellectuals because the appearance of the animals was inconsistent with their existing knowledge.

By delving into the various phenomena brought about by the pair of camels, we will obtain a glimpse of the cultural dynamics surrounding exotic animals in the Edo period. Particularly notable is the fact that a wealth of connotations generated and propagated by the camels found their way into novels, essays, comic verses, popular songs, woodblock prints, and hanging scroll paintings. People's general view of animals is reflected in the image of the camel.

On Exotic Animals

Early documents such as *Chronicles of Japan* (Nihon shoki) and *Continuation of Chronicles of Japan* (Shoku nihongi) state that exotic animals were imported in the late sixth century. In 598, an envoy of the empress Suiko (r. 592–628) brought a pair of magpies from the Silla kingdom back to Japan; moreover, the king of Silla presented the empress with a pair of peacocks as diplomatic gifts. In the following year, the king of Baekje presented the empress with other exotic animals, including a camel, a donkey, two sheep, and a white pheasant. From the seventh through the ninth centuries, rare animals were sporadically imported into Japan as diplomatic gifts from the Korean peninsula.[3]

During the Heian period, when Chinese maritime traders monopolised the export of rare animals to Japan, peacocks and parrots were especially popular. These exotic birds attracted the nobility, including emperors, retired emperors, members of the imperial family, and high-ranking aristocrats in Kyoto; they exchanged the birds among themselves to maintain their social networks. The imported birds in the palace garden and the gardens of noble residences were not just decorative, but also had significant religious meanings. The aristocracy saw peacocks as symbols of an esoteric Buddhist ritual conducted according to the *Mahamayuri Sutra*, or peacock sutra, to avert natural disasters and disease, and to bring happiness. They

2 Asakura Musei, *Misemono kenkyū* (Kyoto: Shibunkaku shuppan, 1977 [1928]), pp. 191–192.

3 Kawazoe Yū, "Hakurai dōbutsu to misemono" (Imported Animals and Shows), in Nakazawa Katsuaki, ed., *Rekishi no naka no dōbutsu tachi* (Animals in History), Hito to dōbutsu no Nihon shi, vol. 2 (Tokyo: Yoshikawa kōbunkan, 2009), pp. 129–130. For the relation between exotic animals and Japanese culture, see Federico Marcon, "All Creatures Great and Small: Tokugawa Japan and Its Animals," in Robert T. Singer and Kawai Masatomo, eds., *The Life of Animals in Japanese Art*, exhibition catalogue (Princeton, NJ: Princeton University Press, 2019), pp. 30–32.

also associated parrots with the preaching of Buddhist teachings because of the birds' ability to echo human speech.[4]

The Sino-Japanese trade reached its climax in the fifteenth century, during the Ming dynasty (1368–1644), but there is little information about the import of exotic animals at that time, apart from a type of hawk native to the Korean peninsula. The Japanese warlords believed that Korean hawks were the best for hunting. In 1613, the first Tokugawa shogun, Ieyasu (1542–1616), ordered the Sō family, whose domain was Tsushima Island, located halfway between the Japanese archipelago and the Korean peninsula, to import hawks from Korea.[5]

Towards the end of the sixteenth century, the scope of the trade expanded to Southeast Asian countries. This meant that more exotic animals native to Southeast Asia reached Japan. In 1594, for example, a merchant in the port city of Sakai presented the hegemon Toyotomi Hideyoshi (1536–1598) with a civet cat he had obtained in the Philippines. Moreover, a series of genre paintings on folding screens from the early seventeenth century, called *Scenes along the Shijō Riverbed*, depict audiences enjoying animal shows of exotic species, such as peacocks, tigers, hedgehogs, and water buffalos, along with kabuki and puppet theatre and acrobat shows, all of which were performed on the dry riverbed of the Kamo River across Shijō avenue in southeast Kyoto.[6]

Issuing the edicts of national seclusion in the 1630s, the Tokugawa shogunate made it impossible for the Japanese to travel abroad or return to their native country in 1635 and barred Portuguese carracks from entering any Japanese port in 1639. After relocating the trading station of the Dutch VOC from Hirado to Nagasaki in 1641, the shogunate restricted their foreign trade partners to three designated countries, the Netherlands, Joseon Korea, and Ming (later Qing) China. As a result, the import of exotic animals was predominated by these countries. A large portion of animal imports were rare birds such as parakeets, parrots, hill myna birds, and pheasants. While some of them were displayed in public shows, others were sold at a high price to domain lords (*daimyō*) and wealthy townsmen.[7]

Among the *daimyō* of the Edo period, there were a few who took a fancy to collecting rare living animals. Typical is Tokugawa Mitsukuni (1628–1700), lord of Mito domain, who was said to keep imported peacocks, parrots, and parakeets, as

4 Kawazoe Yū, "Hakurai dōbutsu to misemono," pp. 131–132.

5 Kawazoe Yū, "Hakurai dōbutsu to misemono," pp. 133–135.

6 Kawazoe Yū, "Hakurai dōbutsu to misemono," pp. 136, 143–144.

7 Kawazoe Yū, "Hakurai dōbutsu to misemono," pp. 136–139. For the reception of exotic birds in the Edo period, see Martha Chaiklin, "Exotic-Bird Collecting in Early-Modern Japan," in Gregory M. Pflugfelder and Brett L. Walker, eds., *JAPANimals: History and Culture in Japan's Animal Life* (Ann Arbor, MI: Center for Japanese Studies, University of Michigan, 2005), pp. 125–160, especially "The Exotic-Bird Trade" and "Avian Acculturation," pp. 139–147.

well as exotic hedgehogs, apes, and civet cats in his residence. The eighth shogun, Tokugawa Yoshimune (1684–1751), enthusiastically imported horses and dogs for official use. Records also show that he ordered elephants from Southeast Asia, and, in 1728, a Chinese ship brought a pair of Indian elephants from Vietnam to Nagasaki. Although the female elephant soon died, the male was kept for over ten years at the shogun's coastal retirement villa called Hama-goten and was sometimes shown to the public.[8]

During this period, the Dutch VOC also presented exotic animals as diplomatic gifts. They brought camels into Nagasaki in 1646 and transported them to Edo for an audience with the shogun in the courtyard of Edo castle in 1647. In 1675, they brought a pair of donkeys from Africa, which were presented at the shogun's court in 1676. In the context of the official audiences with the shogun in Edo, the VOC frequently imported large animals such as elephants, oxen, and horses, as well as exotic birds. They also presented various artworks, wine, food, and other products novel to the shogunal court. These gifts played an important part in maintaining cordial relations with the shogun and his officials, and they were intended to ensure the commercial success of the company in Japan.[9]

This brief history of the import of rare animals into Japan foreshadows the rejection of our pair of camels by the shogun and their subsequent exhibition to the public. In 1821, the VOC presented them to the governor of the city of Nagasaki for the usual inspection. However, the governor refused to accept the camels due to the shogun's disinterest. In his diary, Blomhoff claimed that he offered the camels to a Dutch interpreter, and another document tells us that they were finally obtained by an impresario with the Nagasaki merchant Toyamaya Bunzaemon acting as go-between.[10]

Depicting Camels Before the Edo Period

Visual images of camels appearing in artworks and book illustrations clarify what people knew about these animals. Moreover, these images sometimes suggest the context in which people received the depicted animals.

The earliest examples of camel imagery depict double-humped camels. They appear on two eighth-century objects in the Shōsōin treasury in the precincts of Tōdai-ji temple in Nara: a five-string sandalwood lute inlaid with mother-of-pearl

8 Kawazoe Yū, "Hakurai dōbutsu to misemono," pp. 140–141.

9 Michael Laver, *The Dutch East India Company in Early Modern Japan: Gift Giving and Diplomacy* (London and New York: Bloomsbury Academic, 2020), pp. 37–38, 46.

10 Sugita Hideaki, "Rakuda to Nihonjin: Dōbutsu hyōshō o tōshite mita ikoku shumi" (The Camel in Japanese Literature and Art), *Hikaku bungaku kenkyū* 86 (2009), p. 51.

and a sandalwood chessboard inlaid with wood. One of the earliest paintings with camel imagery is a fourteenth-century hanging scroll titled *Illustrated Biography of the Prince Shōtoku* (Shōtoku taishi eden) in the collection of Shōmankō-ji temple in Aichi prefecture, which features a donkey-like animal with two humps on its back. This strange animal can be safely identified as a Bactrian camel presented by a Korean king to Prince Shōtoku, because the animal is described in *Biography of Prince Shōtoku* (Shōtoku taishi denryaku) attributed to Fujiwara Kanesuke (877–933), the text on which the painting was based. Although no camel appears in the oldest extant work on the same subject from the eleventh century in the collection of Tokyo National Museum, the camel may have been depicted in lost works that faithfully visualised the biography of Prince Shōtoku. Images of double-humped camels may also be found in works depicting the nirvana of the Historical Buddha (*parinirvana*), the earliest examples of which appear in the thirteenth century. Buddhist paintings of this period were generally inspired by the style of imported Song-dynasty paintings.[11] Japanese painters copied the original iconography of double-humped camels, along with many other animals gathering around the Buddha on the bier, without identifying the animal. An example from the fifteenth century is a double-humped camel appearing in the handscroll painting *Figures from Various Countries* (*Kuniguni jinbutsu zukan*), attributed to Sesshū (1420–1506), in the collection of Kyoto National Museum. In the late sixteenth century, more realistic depictions of double-humped camels appear in some *Nanban byōbu*, a pair of folding screens with scenes of trading with the Portuguese, showing the Portuguese, their carracks, and other curios. A sheet of drawings from the late seventeenth century found in the *Sketches of Birds and Animals* by Ogata Kōrin (1658–1716) contains an image of a single-humped camel.[12] The image is presumed to be based on a picture depicting one of the camels, which, as mentioned above, the VOC in Nagasaki imported in 1646 to present to the shogun Tokugawa Iemitsu (1604–1651). For reasons yet unclear, subsequent generations did not make use of the camel imagery that had taken shape by the end of the seventeenth century.

In the eighteenth century, the economic development and the growth of cities created new dynamics in various fields of cultural production including the study of materia medica (*honzōgaku*), originating in Chinese learning, and *Dutch learning*

11 See Nakano Genzō, *Nihonjin no dōbutsu ga: Kodai kara kindai made no ayumi* (Japanese Animal Paintings: From Ancient to Modern Times, Tokyo: Asahi shinbunsha, 1986), pp. 179–180.

12 Belonging to the Ogata Kōrin archives of the Konishi family in the collection of Kyoto National Museum, the sheets of drawings of various birds and animals are mounted into two handscrolls. See Kano Hiroyuki, *Kōrin geijutsu no kisō* (Basic Elements of Kōrin's Art), Nihon no bijutsu 462 (Tokyo: Shibundō, 2004), pl. 2; and Yamane Yūzō, *Konishi-ke kyūzō Kōrin kankei shiryō to sono kenkyū: Shiryō* (The Former Konishi Family's Collection of Kōrin Archives and Its Study, Tokyo: Chūōkōron bijutsu shuppan, 1962), frontispiece 8.

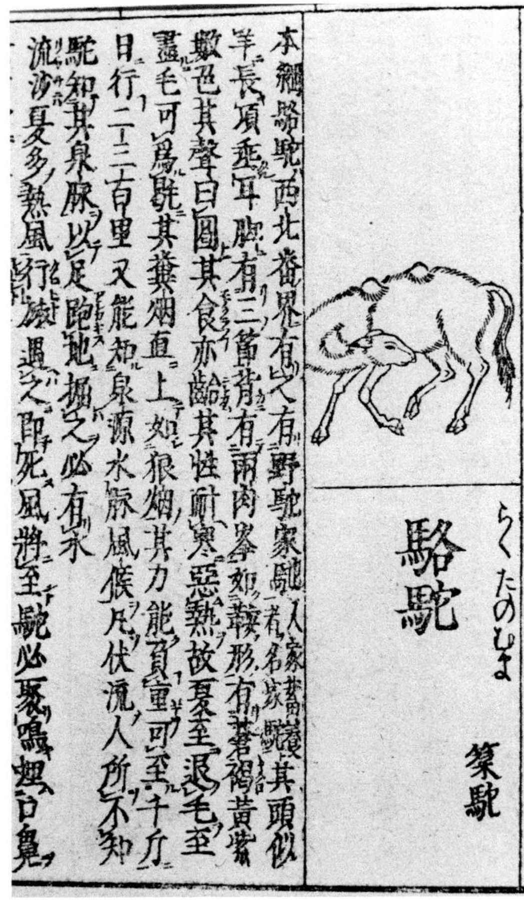

Figure 5.1: *Camel* in Terajima Ryōan, *Illustrations of Three Powers in Japan and China* (*Wakan sansai zue*), vol. 37, 1712. Source: *Wakan sansai zue*, part II (Tokyo: Tokyo bijutsu, 1970, p. 435).

(*rangaku*) imported from the West through the port of Nagasaki. This scientific interest brought camel imagery to a new phase. In general, illustrations in imported Western books broadened the themes and techniques of woodblock prints. For example, the Dutch translation of *Historiae naturalis* by the Polish scholar John Jonston (1603–1675), published in Amsterdam in 1657, contains fine copperplate illustrations depicting various forms of life on earth, including camels. An image of a single-humped camel which is based on a *Historiae naturalis* illustration appears in *Manual of Painting of the Past and Present in Eight Subsequent Varieties* (Kokon gasū kō hasshu), published in 1771 by the Chinese-style painter Sō Shiseki (1715–1786).[13] However, the rarity of imported books from the West may be the reason that scholars did not mention the camels illustrated in *Historiae naturalis* when they discussed the animal after the arrival of our one-humped camels in 1821,

13 For Sō Shiseki's camel illustration, see Marcon, "All Creatures Great and Small," p. 36.

whereas images from popular illustrated books of Chinese learning were used in the discussion. One example is a small illustration of a double-humped camel appearing in the illustrated Sino-Japanese encyclopaedia *Illustrations of Three Powers in Japan and China* (Wakan sansai zue), compiled by Terajima Ryōan (dates unknown) and published in 1712 (Figure 5.1). The Ming-dynasty compendium of medical materials *Compendium of Materia Medica* (Chin. *Bencao gangmu*, Jap. *Honzō kōmoku,* 1596) by Li Shizhen (1518–1593) was frequently referred to as an authentic source of knowledge about camels, even though it does not contain any illustrations of the animal in question. Another source was provided by a double-humped camel that an American ship had carried to Nagasaki in 1803. Although the camel was not allowed to disembark, information about the camel found its way into the country because of the publication of woodblock prints depicting the animal.

The Popular Reception and Critical Eyes on Camels

Woodblock prints published in Nagasaki, commonly known as 'Nagasaki prints,' contributed to popularising the pair of camels, as was the case for other exotic animals and birds imported to the port city, and, as a result, contributed to the success of the camel show. Combining camels, Dutchmen and their enslaved attendants in a single image, Nagasaki prints not only visualised the dimensions of the camels, but also highlighted the exotic atmosphere of the strange animals imported into Nagasaki from a remote corner of the world. Some prints bear captions describing the animal in minute detail, for example with the title, "The Dutch imports early in the seventh month of the fourth year of the Bunsei Era [1821]" (Figure 5.2), continuing:

> A pair of camels, the male being four years old, the female five. Their approximate length is one *jō* and five *shaku* [fifteen feet], and their height nine *shaku* [nine feet]. Camels of this kind are reared by farming families in that country [Persia] and serve in farming. When used as a draught animal, it covers one hundred *ri* [about 250 miles] in a day without getting tired. When they are loaded, they fold their legs in three. First, they fold their front legs so that it is easy to put the load on their backs. They can stand up with a load as heavy as one hundred *kin* [about 130 pounds]. The male and the female are very affectionate towards each other. They are gentle animals.[14]

These captions on the prints provided people with general information about the animal, however, it was often incorrect. Moreover, those who made a living from

14 *Camels*, Nagasaki print, 1821. Higuchi Hiroshi, *Nagasaki ukiyo-e* (Nagasaki prints, Tokyo: Mitō sho'oku, 1871), pl. 8.

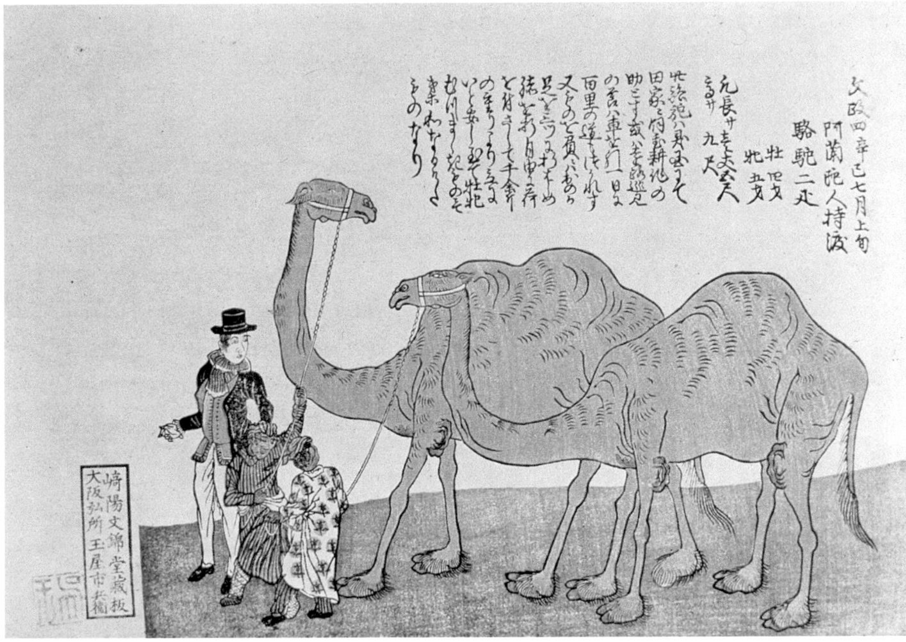

Figure 5.2: *Camels*, Nagasaki print, 1821. Source: Higuchi Hiroshi, *Nagasaki ukiyo-e* (Tokyo: Mitō sho'oku, 1871, pl. 8).

the camel shows appropriated the captions for their business, and the animals' character tended to be exaggerated as the shows became popular.

Our pair of single-humped camels were exhibited in Osaka and Kyoto in 1823, in Edo in the following year, and in Nagoya in 1826. In *A Study of Public Shows*, Asakura Musei summarises the programme of the shows as follows:

> When the spectators have all gathered in the barn where the show is held, a presenter [*kōjō*] in Chinese costume comes out, leading the camels around a circular space while a band plays the *shamisen*, bell, drum, and flute. He finally leads them to the centre of the circle and starts to speak about the strange animals. Following the presenter, a servant feeds the camels some vegetables, such as *daikon* radishes and sweet potatoes, while a vendor with vegetables in a basket walks around the circle to sell them to the spectators. Spectators who buy some [vegetables] enjoy feeding them to the animals. At the end of the programme, the camels slowly walk around the circle twice, led by three men in Chinese costume, to the music of the band. When a man beats a drum to announce the end of the show, the spectators leave through the gate.[15]

15 Asakura Musei, *Misemono kenkyū*, pp. 196–197.

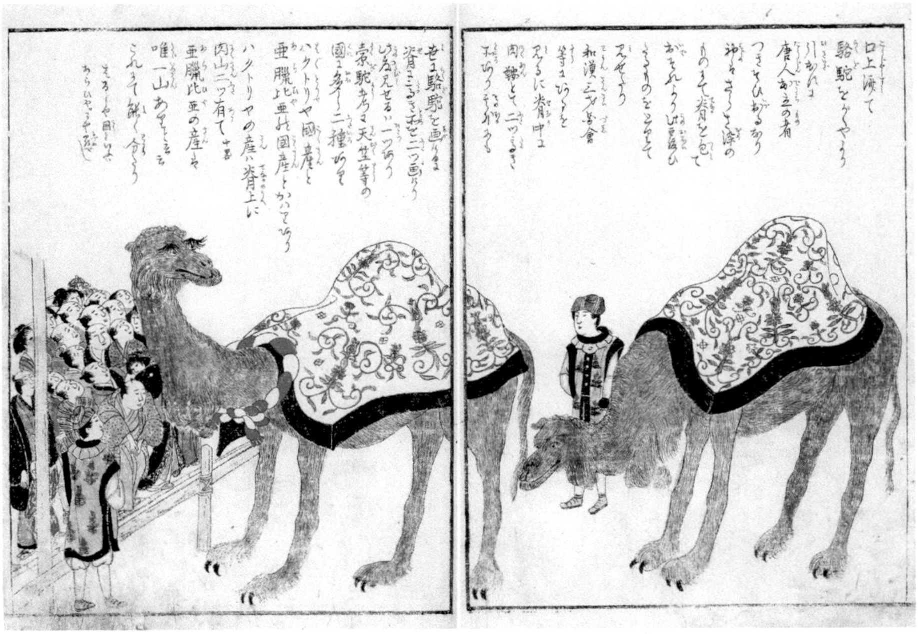

Figure 5.3: Kōriki Enkōan, *An Illustrated and Detailed Record of Camels* (*Ehon rakuda gushi*), manuscript, 1827, Nagoya City Museum. Source: *An Illustrated and Detailed Record of Camels* (*Ehon rakuda gushi*). *Nagoya-shi hakubutsukan shiryō sōsho* 3 (Nagoya: Nagoya-shi hakubutsukan, 2007, pp. 18-19).

While Asakura does not mention the 'stage setting' of the show, the manuscript *An Illustrated and Detailed Record of Camels* (Ehon rakuda gushi, 1827) by the Nagoya essayist Kōriki Enkōan (1756–1831) visualises the spectacle with a few illustrations by his own hand (Figure 5.3). Describing the camel show in his home city, Enkōan's work testifies to the popularity of the camels. The manuscript, for example, visually illustrates camel-show-related merchandise such as clay figures, a *sugoroku* game, dolls, kites, a fan, and a tobacco pouch.[16]

The camel shows generated various connotations, such as the idea of 'a harmonious couple.' A caption of a handbill (*hikifuda*) for one of the camel shows (Figure 5.4) reads:

> The female and the male are both gentle-tempered. Nothing is more harmonious than this pair of camels, so that a human couple who see the pair will acquire the animals' capacity and become harmonious as well.[17]

16 See *Ehon rakuda gushi*, ed. Nagoya-shi hakubutsukan, in *Nagoya-shi hakubutsukan shiryō sōsho* (Series of Materials in the Collection of Nagoya City Museum), vol. 3 (Nagoya: Nagoya-shi hakubutsukan, 2007), pp. 37–38, 42, 44, 54–56.

17 *Camels*, flier (Kyoto: Yoshinoya Kanbei; and Osaka: Jun'idō; Tamaya Ichibei, ca. 1823), in the collection of the National Museum of Japanese History.

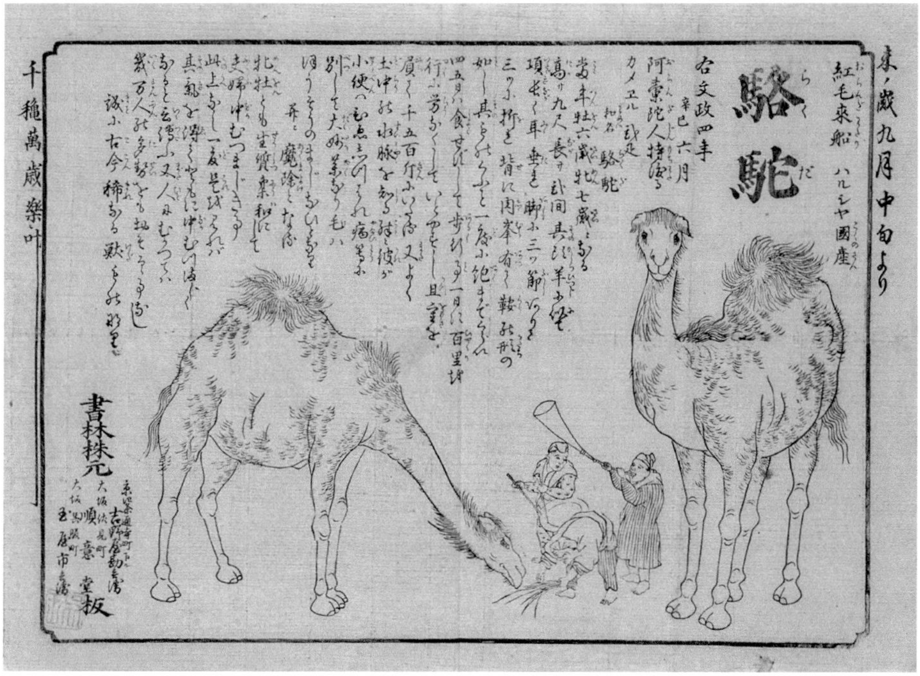

Figure 5.4: *Camels*, Handbill, woodblock print on paper, 34.5 x 48.0 cm, Kyoto: Yoshinoya Kanbei; Osaka: Jun'idō; Osaka: Tamaya Ichibei, ca. 1823, National Museum of Japanese History.

The flier also mentions more practical benefits, for instance, that camel hair can be used as an amulet against smallpox and to ward off evil in general. These magical functions had no basis in any knowledge of natural history but were fabricated by the presenters of the camel shows. The fliers were widely circulated and attracted people's attention.

The Japanese word for camel, *rakuda*, originally a Chinese term, now came to entail the idea of 'a harmonious couple' in everyday life. The Osaka novelist and *ukiyo-e* painter, Akatsuki Kanenari (1793–1861), commented on the popularity of the term:

> Nothing is more laughable than the recent use of a trendy word. Since the abovementioned pair of camels were imported, the word *rakuda* has become the fashionable term for a married couple who go walking together.[18]

18 Akatsuki Kanenari, *Unkin zuihitsu* (Essay on Beautiful Morning Mist), vol. 4 (1862), in *Nihon zuihitsu taisei* (Compendia of Essays in Japan), ed. Nihon zuihitsu taisei henshūbu, series 1, vol. 3 (Tokyo: Yoshikawa kōbunkan, 1975), p. 113.

At the same time, *rakuda* connoted a quality of uselessness due to the camels' sluggish movement in the shows. According to Shihekian Shigetsuta's (dates unknown, active in the nineteenth century) observation,

> At that time a pair of large animals were imported from Persia, and they were called *rakuda* […]. It is said that in that country camels are made useful by carrying heavy loads on their backs and travelling great distances. In this country, however, people do not know how to use them. Because of this, people call what is large and useless *rakuda*.[19]

In addition to conjugal harmony and uselessness, urban dwellers discovered another characteristic of the camel: it was supposed to be 'easy-going.' No doubt, it was the term *rakuda* that suggested this quality: *raku* evokes a sense of leisurely existence or the state of being carefree, while the suffix *da* is a verb signifying 'being.' Therefore, camels came to be associated with such a carefree state of existence. In *Annals of Edo* (Bukō nenpyō, 1850), the eminent writer Saitō Gesshin (1804–1878) recorded a comic verse (*kyōka*) punning on *rakuda* composed by the Kyoto poet Kamo Suetaka (1752–1841):

Kubi wa tsuru	With a neck resembling a crane's
senaka wa kame no	and a back the shell of a tortoise
kō ni nite	easy-going are camels
senshū rakuda	for a thousand years
banzei rakuda	for ten thousand years[20]

In a collection of bits of hearsay from Edo, *Trivial Talks of the Town* (Kōgai zeisetsu, 1829), the author Jinsaiō (dates unknown) recorded a popular song making fun of the pair of camels:

Anoya, Harusha-koku de wa	Ah, in Persia
ichi-nichi ni sen-ri mo	they used to walk as far as
aruite mita ga eh	one thousand *ri* a day
Edo ja kuccha necha	in Edo they only eat and sleep
kuccha necha shicha	and only do eat and sleep

19 Shihekian Shigetsuta, *Wasure nokori* (Remains Left Behind), vol. 2 (1824), in *Zoku enseki jisshu* (Ten Kinds of Stones Looking Like Jade, second series), vol. 2 (Tokyo: Chūō kōronsha, 1980), pp. 148–149.

20 Entry for the sixth month of Bunsei 4 (1821), Saitō Gesshin, *Bukō nenpyō* (1850). See *Zōho bukō nenpyō* (Annals of Edo, enlarged edition), ed. Kaneko Mitsuharu, vol. 2, Tōyō bunko 118 (Tokyo: Heibonsha, 1968), p. 68.

koitsu wa mata	this seems to be
rakuda rō	only easy-going[21]

Whereas in the caption to the Nagasaki print the camel walks a hundred *ri* a day in their native country, this song speaks of a thousand *ri*. By sharpening the contrast, the song underlines the animals' easy-going existence in Edo.

Multiple meanings of *rakuda* coexisted in the popular imagination. One notable example is *The Harmonious World of Camels* (Wagō rakuda no sekai, 1825), an illustrated popular novel (*gōkan*), in which the author Kōnantei Karatachi (dates unknown) tells the story of a married couple who learn from the animal how to lead a happy life. The author appropriated the currently popular symbol of the 'harmonious couple' to advocate the feudal morality of conjugal harmony. The husband and wife live in a tenement house and, in one scene (Figure 5.5), their landlord, Santarō, settles a quarrel between them by giving them a lecture on camels:

> The pair of camels [...] they're such a harmonious couple. Even creatures of this kind act like this. The female cares for the male, and the male cares for the female [...]. Anyway, as a human couple, try to be harmonious and work hard day and night. Then your business will flourish, and naturally your descendants will be happier. Don't quarrel with each other anymore [...]. Work together to make money with unflagging zeal. Then you'll be easy-going [*rakuda*] in your future life.[22]

The appearance and popularity of the pair of camels in the show likely inspired the author of a popular novel to present the camels as an ideal couple exemplifying conjugal unity, happiness, and prosperity.

The camel show also attracted the attention of the intelligentsia. Having observed the camels with their own eyes, learned people of the time began to revisit long-accepted views of the animal. In the field of natural history, Dutch-learning scholars fiercely attacked the views of their Chinese-learning competitors. In 1824, Ōtsuki Gentaku (1757–1827), an eminent scholar of Dutch learning, challenged the conventional understanding of camels voiced two decades earlier by Ono Ranzan (1729–1810), a scholar of Chinese learning:

> In *Notes on What I Have Heard* (Kibun), Ono Ranzan says that the body [of a camel] is large and its shape mostly resembles a sheep [...]. He just had this from

21 Jinsaiō, *Kōgai zeisetsu*, vol. 2 (1829), in *Zoku Nihon zuihitsu taisei* (Compendia of Essays in Japan, second series), eds. Mori Senzō and Kitagawa Hirokuni, supplement 9 (Tokyo: Yoshikawa kōbunkan, 1983), p. 131. One *ri* is 2.44 miles.

22 Kōnantei Karatachi, *Wagō rakuda no sekai*, illustrated by Utagawa Kuniyasu (Edo: Moriya Jihei, 1825), 8 verso–9 recto.

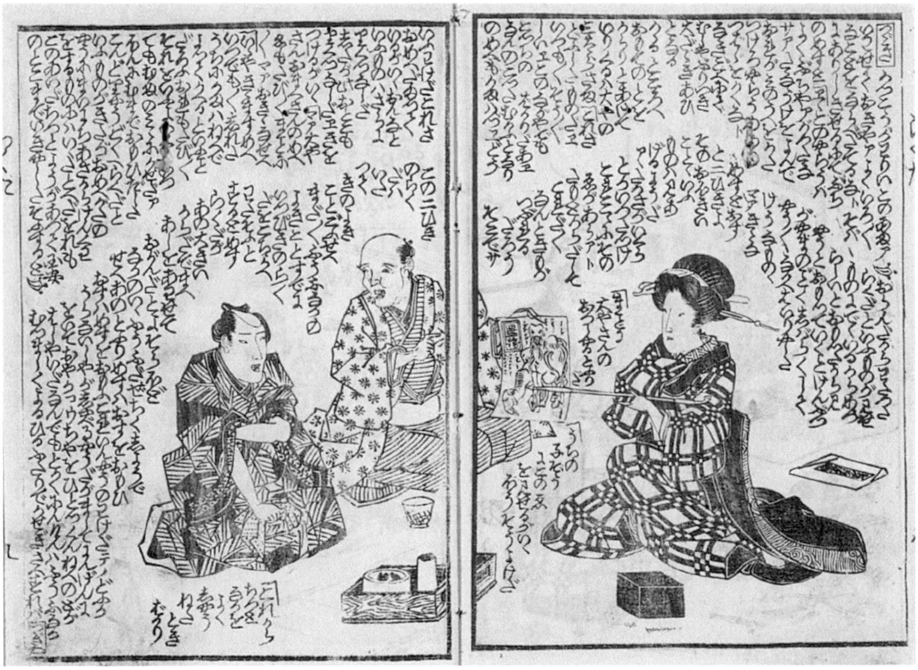

Figure 5.5: Kōnantei Karatachi, *The Harmonious World of Camels* (*Wagō rakuda no sekai*), illustrated novel. Edo: Moriya Jihei, 1825, 8 verso and 9 recto, National Diet Library, Tokyo.

> hearsay from the time [when a double-humped camel was imported in 1803]. Probably all the past and present views [on camels] in books of Chinese learning were just based on hearsay. If so, it is only natural that they cannot clarify the original habitat of camels.[23]

Ōtsuki's criticism of Ono Ranzan highlighted the lack of empiricism in the approaches of Chinese-learning scholars. This gave the impression that the knowledge of Chinese-learning scholars was unreliable and, in this way, elevated the status of scholars of Dutch learning.

Many intellectuals of the time tended to favour the conclusions of Dutch learning because of the empirical attitude of the discipline. They were particularly concerned about the disparity between the idea that was presented in existing literature regarding camels and the actual appearance of the animals they observed in the show. The bone of contention was the number of humps. The camels they saw were single-humped, while those in the literature were often double-humped. In *Annals*

23 Ōtsuki Gentaku, "Rakuda mondō" (Dialogue on camels, 1824), in *Ran'en tekihō* (Gathering flowers in an orchid field, 1831), manuscript in the collection of Tokyo National Museum.

of Edo, apparently recognising the point, Saitō Gesshin criticised the Sino-Japanese encyclopaedia *Illustrations of Three Powers in Japan and China*:

> Seeing the genuine things [camels] this time, I realised that the illustration Tachibana Morikuni drew in *Illustrations of Three Powers in Japan and China* is false. He drew two humps on the back [of the animal] based on the view that the humps on its back look like a saddle. In reality, it has only one hump. Moreover, the hump is very high.[24]

In his *Trivial Talks of the Town*, the abovementioned Jinsaiō echoes Saitō Gesshin's criticism:

> The statement that the humps are one behind the other and have a saddle-like shape is false. [And in reality] the hair resembles that of bovines and the colour is like that of "red cattle" (*aka ushi*). And it smells like cattle and is a sluggish animal. The male has thick front legs and lots of black hair covering its eyes. One cannot tell whether these are eyebrows or eyelashes. It moves back and forth in a curious rhythm to the tune of a flute and a drum.[25]

The discrepancy between the image based on the observation of the camels and the knowledge provided by the encyclopaedia clearly caused confusion among the intellectuals. However, the empiricism of the time soon led them to the recognition of another kind of camel with a single hump.[26] The journal of the retired lord of Hirado Domain in Kyūshū, Matsura Seizan (1760–1841), *Evening Tales of Months and Years Past* (Kasshi yawa), shows the process by which he came to know the animal, correcting his knowledge step by step. The process started when he encountered two images, one showing a double-humped camel and the other a single-humped one. In an entry for 1822, Seizan writes:

> A Dutch ship came to Nagasaki last year with camels on board. People said that they would later come to Edo, but in the end they did not. A certain painter gave me a picture of a camel some years ago when I attended a gathering at the residence of a lord. I found the picture among many old papers, and I have put it on the following page.[27]

24 Kaneko Mitsuharu, *Zōho bukō nenpyō*, vol. 2, p. 68.

25 Jinsaiō, *Kōgai zeisetsu*, vol. 1 (1829), in *Zoku Nihon zuihitsu taisei*, supplement 9, p. 130.

26 In *Bencao gangmu*, Li Shizhen refers to single-humped camels inhabiting Turfan (present-day eastern part of Xinjiang Uygur, China), but scholars seem to have forgotten about it until they recognised single-humped camels.

27 Matsura Seizan, "Entry for Bunsei 5 (1822)," in *Kasshi yawa*, vol. 8, reprinted in Nakamura Yukihiko and Nakano Mitsutoshi eds., *Kasshi yawa*, vol. 1, Tōyō bunko 306 (Tokyo: Heibonsha, 1977), p. 136.

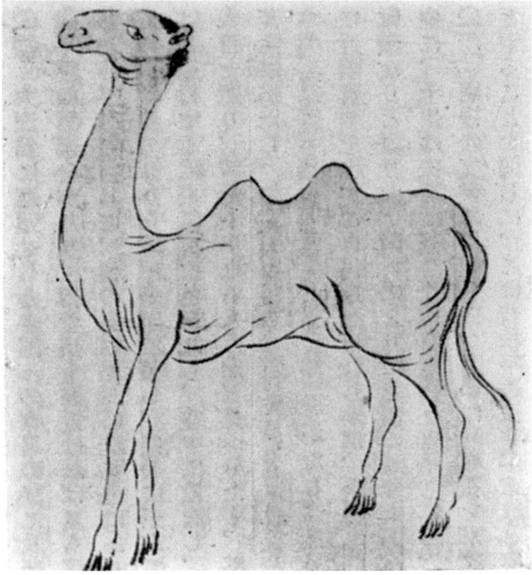

Figure 5.6: *Camel* in Matsura Seizan, *Evening Tales of Months and Years Past (Kasshi yawa)*, vol. 8, 1822. Source: *Kasshi yawa* 1, Tōyō bunko 306 (Tokyo: Heibonsha, 1977, p. 136).

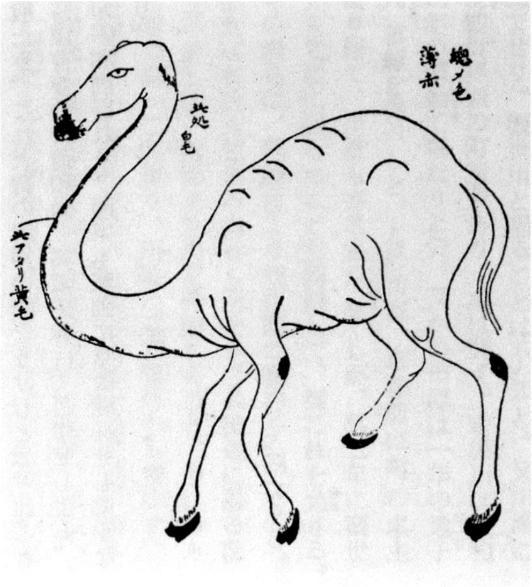

Figure 5.7: *Camel* in Matsura Seizan, *Evening Tales of Months and Years Past (Kasshi yawa)*, vol. 9, 1822. Source: *Kasshi yawa* 1, Tōyō bunko 306 (Tokyo: Heibonsha, 1977, p. 163).

The image this entry refers to depicts a camel with two humps, also called saddle-humped (Figure 5.6). Another picture that Seizan saw in 1822, after having written this entry, depicts a single-humped camel that was said to come to Edo soon. This picture (Figure 5.7) considerably confused him:

> I happened to see a signboard of a show when passing by Ryōgoku bridge in the third month. It illustrated camels, and they sold woodblock prints depicting the animal [...]. I sent a man to ask what the picture was about, and they answered that it was an image of the camels brought to Nagasaki last year, and that the genuine thing would soon come to this city. The following day, I sent a man to see [what] the show [would be about]. They displayed a *tsukuri-mono* [a stage prop emulating the original figure] for the show, and he took a picture of the prop and came back with it. Judging from the picture, the figure must be a fake, and not modelled after the real animal [...]. This prop, however, does not have the saddle-like humps, which contradicts the caption of the woodblock prints [depicting camels]. The caption corresponds to what I have mentioned about the camel in the previous [eighth] volume of my journal.[28]

28 Matsura Seizan, "Entry for Bunsei 5 (1822)," in *Kasshi yawa*, vol. 8, reprinted in Nakamura Yukihiko and Nakano Mitsutoshi eds., *Kasshi yawa*, vol. 1, Tōyō bunko 306 (Tokyo: Heibonsha, 1977), pp. 163-164.

It all became clear when the real camels came to Edo two years later, in 1824 (Bunsei 7), and Seizan finally became aware of the existence of another kind of camel with a single hump:

> This year the camels came from Nagasaki to this capital [Edo] [...]. There are different kinds of camels. This time it is a kind that is called single-humped camel. The prop I mentioned earlier [in volume nine] was indeed of this kind. The camel recorded in volume eight was of a kind that had arrived in the seventh month of the third year of the Kyōwa era [1803] [...]. This is definitely a different kind of camel because the so-called saddle-like humps are depicted in the picture.[29]

Matsura Seizan's experience is representative of the way in which knowledge about camels spread to various social groups who were, each in their own way, curious about unknown creatures. It shows that the educated public, people like Gesshin and Seizan, also used *misemono* spectacles as a source of knowledge. Seizan's progress seems to be part of a general interest in natural objects shared by people of the time.

Commissioned Camel Paintings

Two extant works by the eminent painters Tani Bunchō (1763–1840) in Edo and Maruyama Ōshin (1790-1838) in Kyoto suggest that they were commissioned to paint the pair of camels imported in 1821; nothing, however, is known about the clients.[30] The demand for camel paintings was so large that the minor painter Kubota Setsuyō (dates unknown), who was based in Kyoto, repeatedly used the same composition to paint the animals. Three extant paintings of camels by Setsuyō are commonly dated to the tenth month of 1823,[31] the one by Ōshin to the ninth

29 Matsura Seizan, "Entry for Bunsei 7 (1824)," in *Kasshi yawa*, vol. 53, reprinted in Nakamura Yukihiko and Nakano Mitsutoshi eds., *Kasshi yawa*, vol. 4, Tōyō bunko 333 (Tokyo: Heibonsha, 1978), p. 73.

30 The most recent work mentioning Maruyma Ōshin's *Camels* (Figure 5.8), as well as other camel images from the eighth century to Hokusai's works, is Ayelet Zohar, *The Curious Case of the Camel in Modern Japan: (De)colonialism, Orientalism, and Imagining Asia* (Leiden: Brill, 2022), in which she analyses Ōshin's work from a viewpoint of the realistic style characterising paintings by the Maruyama school. See "*Shashin*, Realism, and Live Drawings of Camels" in Chapter 2, pp. 73-79.

31 See *Misemono: Spectacles, Shows, and Circuses in the Edo-Meiji Period*, exhibition catalogue (Tokyo: Tobacco and Salt Museum, 2003), p. 58; *Cute Edo Paintings*, exhibition catalogue (Fuchu: Fuchu Art Museum, 2013), p. 44; and Fuchu Art Museum ed., *Dōbutsu no e: Nihon to Yōroppa, fushigi kawaii heso magari* (Tokyo: Kōdansha, 2021), p. 141. This catalogue contains another painting of *Camels* (1823) by Ueda Kōchō (1788-1850) in a private collection.

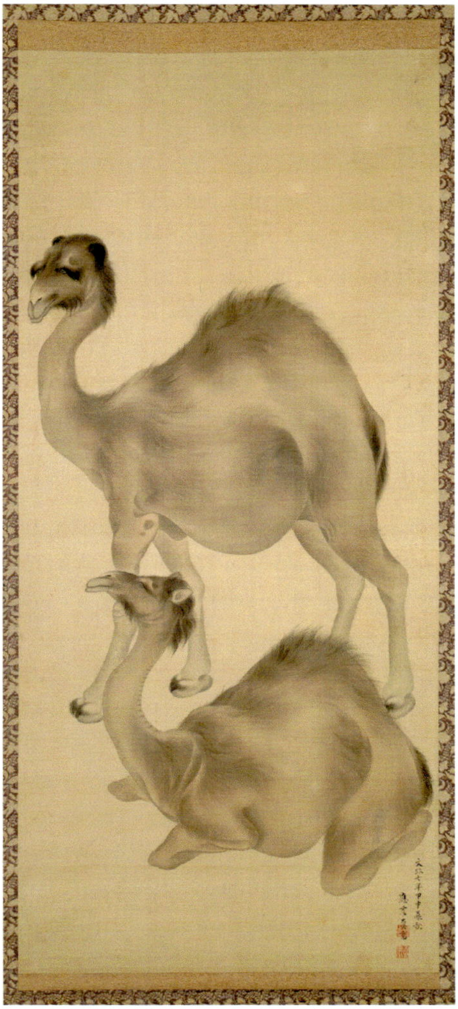

Figure 5.8: Maruyama Ōshin, *Camels*, hanging scroll, colour on silk, 125.5 x 54.6 cm, 1824, the Etsuko and Joe Price Collection.

month of 1824 (Figure 5.8), and the one by Bunchō to the twelfth month of 1825.[32] Setsuyō's inscription indicates that he painted the camel from life. It is possible that he made some drawings at a camel show in Kyoto, which he later used to supply paintings for commissions. The paintings by Ōshin and Bunchō were likewise made when the show was taking place in the painters' home cities, Kyoto and Edo. Their production dates suggest that Ōshin and Bunchō painted the strange animals after observing the animals at the show. Although we cannot tell whether they saw the

32 For Bunchō's *Camels* (hanging scroll, colour on silk, 98.0 x 42.0 cm., in the collection of Tekisuiken Cultural Foundation), see *Tani Bunchō: Commemorating the 250th Anniversary of His Birth*, exhibition catalogue (Tokyo: Suntory Museum of Art, 2013), p. 26.

camels before receiving the commissions or after, their naturalistic painting skills would have enabled them to meet the expectations of the clients.

Realistic depictions of this kind suggest a genuine interest in the actual animal rather than in its symbolic or talismanic connotations as highlighted by other media such as handbills distributed in connection with the camel shows. Their styles reveal subtle differences in the various painting techniques applied in these works, such as the use of ink wash and shading, but the painters shared a common attitude towards their subject: they aimed at a life-like representation of the animal. It should be noted that the three painters consciously chose to leave the background of their composition blank, in strong contrast to the Nagasaki prints which depict the camels within an elaborate setting. The absence of environment results in a portrait-like impression, focusing the viewer's attention on the individual shape and appearance of the animal. All three painters probably judged that this compositional style most suited the requirements of their clients.

In early modern East Asia, an empty background put the focus on the individuality of the subject, particularly in ancestral portraits. For this reason, a portrait usually included attributes that indicated the social status of the sitter, such as a gold-brocade garment for a high-ranking Buddhist monk, to identify the sitter and to highlight their distinguished personality. Similarly, East Asian painters traditionally chose this portrait-like compositional style, leaving the background blank, when depicting individual animals. In reference to extant works, this tradition traces back to the Northern Song dynasty in eleventh-century China, and to the Kamakura period in thirteenth-century Japan.[33] Artists who depicted camels must have been aware of this tradition. Therefore, it is significant that these camel portraits do not include any references to specific social contexts. This suggests that both the clients and the painters were primarily interested in capturing the spirit of the animal, not in its role within a specific context.

Although these painters adopted the traditional background-less style to highlight the individuality of their subject, in a way resembling works of portraiture, it should be noted that the composition and the realistic depiction were not adopted to merely represent a specimen of a particular species of camel. In other words, these painted camels were regarded as specific individuals imported into Nagasaki to 1821, and as such different from other camels belonging to the same species.

The three painters adopted the traditional blank-background style of composition to meet their clients' requirements. Moreover, the painters regarded the composition

33 This painting tradition can be traced back to a handscroll of eleventh-century Northern Song China, *Five Horses* by Li Gonglin in the collection of Tokyo National Museum, and a handscroll of thirteenth-century Japan, *Shungyū-zu* (*Ten Fast Bulls*), eight fragments of which are respectively housed in Tokyo National Museum, Cleveland Museum of Art, Seattle Art Museum and other collections. For the Li Gonglin work, see Itakura Masaaki, *Ri Kōrin* Goba zu (Li Gonglin *Five horses*, Tokyo: Hatori shoten, 2019).

as a kind of framework to depict their subjects as individuals, but not in the same way as one would depict a person in a portrait or flora and fauna as part of a picture. From an art historical point of view, their camels stood midway between a sitter and an illustrated object in a picture book for the study of natural history.

A careful reading of Ōshin's *Camels* and its social context can provide a better understanding of how realistic paintings such as these were appreciated in the late Edo period based on the skill in the painter. In the signature, the painter stresses his skills of realistic depiction: "Drawn from life [*shinsha*] by Ōshin in the ninth month of the seventh year of the Bunsei era."[34] Ōshin was a grandson of Maruyama Ōkyo (1733–1795), the founder of the Maruyama school in Kyoto. Ōkyo was known as the pioneer of naturalistic painting and, accordingly, Ōshin was regarded as a master painter and successor to his grandfather's painting style.

Compared with other surviving camel paintings, Ōshin's stands out for its use of a wide range of painting techniques, such as the delicate use of gradation to depict the round belly of the animals, and the skilful use of ink wash to represent soft fur. Another difference is in how the camels are arranged: Ōshin depicts one camel standing and the other crouching, while both are depicted standing in other works. Probably, the one standing is the male and the recumbent one the female. Ōshin must have adopted contrasting postures to show the curious way in which camels fold their legs when lowering themselves. This was frequently described in the captions that were added to the imagery in handbills and Nagasaki prints.

Close observation of Ōshin's work points to another striking feature. It gives viewers a lively impression of the animals, very different from that given, for example, by a stuffed specimen. This seems to be the result of the painter's intention of giving exotic animals a personality, which is particularly evident in the facial expression of the standing camel with its long black eyebrows or eyelashes. The camel's face must have reminded most viewers of the facial expression of an old man. Of course, the blank background, mentioned above, highlights this personifying effect.

This attempt at humanisation was linked to the notion of "living things" (*shōrui*). *Shōrui* does not distinguish between human beings and animals because it is tied to the Buddhist concept "transmigration to the Six Realms" (*rikudō-rin'ne*).[35] According to this concept, a human being or animal may become any other creature

34 The original signature reads "Bunsei shichi nen kōshin boshū Ōshin shinsha" with two of the painter's seals. For a discussion of the concept *shashin* ("representing the real") in *materia medica* or *honzōgaku* during the late Edo period, see Maki Fukuoka, *The Premise of Fidelity: Science, Visuality, and Representing the Real in Nineteenth-Century Japan* (Stanford: Stanford University Press, 2012), pp. 105-153.

35 For the notion of *shōrui*, see Tsukamoto Manabu, "Mushi o miru me no rekishi: Edo jidai jin o chūshin ni" (A History of the View to Insects: Focusing on People in the Edo Period), *Shakaishi kenkyū* 6 (1985), pp. 71-77.

after death. Some of those who saw the show would have thought that a married couple had become this pair of camels. It also seems natural that people of the time credited the animals with human qualities and feelings. Consequently, this shared notion of *shōrui* may have provided the painter with an incentive to personify the animals in this way. The portrait-like composition apparently fitted this purpose. Surviving works by the two other painters, Bunchō and Setsuyō, may also reflect the same notion of *shōrui*, judging from their realistic painting styles, although they did not have Ōshin's skill in personifying living animals.

Despite Ōshin's statement that he "drew from life", the painter did not try to achieve what we would identify as a realistic or photographic style. He tried to capture the living spirit of the animal following the notion of *shōrui*. In addition, even though contemporary viewers were probably unaware of this, their appreciation of the images of the camels was impacted by the same sense of being alive. When interpreting the painter's signature in this way, one gets a sense of how people at that time may have understood a true-to-life painting of animals.

Comparison with the Case of Elephants

Comparing camels and elephants, two of the largest animals imported into Japan, makes us aware of different routes of reception and representation. Already in the sixteenth century, Portuguese carracks carried elephants along with double-humped camels and other curios to Japan, as can be seen in littoral scenes of the abovementioned screen paintings called *Nanban byōbu*. During the Edo period, elephants were brought to Japan at least three times. The first recorded import concerned a pair of elephants brought by a Chinese ship to Nagasaki in 1728 on the order of Shogun Tokugawa Yoshimune, as mentioned above. The second case was a female elephant from Ceylon brought by the English East India Company to Nagasaki in 1813. Wishing to safeguard Dutch trade interests, the *opperhoofd* of the Dutch factory at Deshima prevented the English from presenting the animal to the shogun. As a result, the elephant was sent back three months later. The third and last case was an Indian elephant taken to Yokohama by an American ship in 1862, eight years after the Tokugawa shogunate and the United States had concluded a treaty of peace and amity, which was followed by treaties with other Western countries. In 1863, the elephant was exhibited in a public show in the amusement district at the west end of Ryōgoku bridge in Edo and subsequently travelled around the country for more than ten years. The camels of 1821 remained the top attraction while the elephant show is said to have attracted the second largest audience.[36]

36 Tobacco and Salt Museum, *Misemono*, pp. 60-61.

Because the show was so popular, a much larger number of *ukiyo-e* prints were produced depicting the elephant of 1862 than those imported earlier. One of the works, a pair of *ōban* size colour prints by Utagawa Yoshitoyo (dates unknown), has a caption with an interesting phrase, which reads, "those who behold it [the elephant] once, will avoid seven kinds of calamity and enjoy seven kinds of happiness."[37] Clearly, as in the case of the camels of 1821, the elephant show propagated the belief that the animal possessed miraculous powers to grant divine favours to the audience. However, there is no evidence that portrait-like paintings of the elephant were commissioned. The most probable reason for this was the dominance of the Buddhist iconography of the elephant. A white elephant was widely recognised as the vehicle of Bodhisattva Samantabhadra. More commonly, on the day of the Buddha's annual memorial service, elephants were seen in paintings of the *parinirvana* along with many other creatures gathering around the Buddha on the bier. Elephants had inherent religious meanings; camels did not. A double-humped camel occasionally appeared in *parinirvana* scenes, as mentioned above, but most people would have been unable to identify it as a camel. The lack of religious connotations allowed people to furnish camels with arbitrary character traits. When commissioned to depict a camel, painters could only draw from life as there was no existing iconography, apart from inaccurate illustrations in encyclopaedias. This situation was favourable to painters skilled in drawing from life, such as Maruyama Ōshin in Kyoto and Tani Bunchō in Edo.

Conclusion

Examining the remarkable ways in which the pair of camels of 1821 were received and represented shines a light onto a broad range of cultural phenomena which would otherwise remain hidden. It gives an insight into the interplay of people's spiritual life, current feudal morality, Buddhist ideas, superstitious beliefs, and popular culture. It demonstrates a shared enthusiasm for curious things that engaged people of every class: scholars, novelists, essayists, and artists. This enthusiasm was also propelled by visual media and public *misemono* shows. Media such as *ukiyo-e* colour prints, monochrome broadsheets, and handbills were instrumental in circulating knowledge, both visually and textually, about these animals. At the same time, the camel shows gratified people's desire to see the living animals by creating a space for encounter. While media provided accurate as well as fabricated information, the spectacle of viewing the animals also engendered emotions, impressions, and connotations. These intersecting processes allow us to observe

37 Tobacco and Salt Museum, *Misemono*, p. 61.

how the cultural dynamics of that time stimulated imagery, essays, novels, songs, verses, and critical engagement.

Glossary

Akatsuki Kanenari	暁鐘成
Aka ushi	赤牛
Asakura Musei	朝倉無聲
Bencao gangmu	本草綱目
Bukō nenpyō	武江年表
Bunsei	文政
Bunsei shichi nen kōshin boshū Ōshin shinsha	文政七年甲申暮秋應震真寫
Ehon rakuda gushi	絵本駱駝具誌
Fujiwara Kanesuke	藤原兼輔
Gōkan	合巻
Hama-goten	濱御殿
Hiki-fuda	引札
Honzōgaku	本草学
Jinsaiō	塵哉翁
Jō	丈
Kamo Suetaka	賀茂季鷹
Kasshi yawa	甲子夜話
Kibun	紀聞
Kin	斤
Kōgai zeisetsu	巷街贅説
Kōjō	口上
Kokon gasū kō hasshu	古今畫藪後八種
Kōnantei Karatachi	江南亭唐立
Kōriki Enkōan	高力猿猴庵
Kubota Setsuyō	窪田雪鷹
Kuniguni jinbutsu zukan	国々人物図巻
Kyōka	狂歌
Kyōwa	享和
Li Gonglin	李公麟
Li Shizhen	李時珍
Maruyama Ōkyo	圓山應擧
Maruyama Ōshin	圓山應震
Matsura Seizan	松浦靜山
Misemono kenkyū	見世物研究

Nagasaki ukiyo-e	長崎浮世絵
Nagoya-shi hakubutsukan shiryō sōsho	名古屋市博物館資料叢書
Nanban byōbu	南蠻屏風
Nihon shoki	日本書紀
Nihon zuihitsu taisei	日本隨筆大成
Ōban	大判
Ogata Kōrin	尾形光琳
Ono Ranzan	小野蘭山
Ōtsuki Gentaku	大槻玄澤
Rakuda	駱駝 /楽だ
Rakuda mondō	駱駝問答
Ran'en tekihō	蘭畹摘芳
Rangaku	蘭学
Ri	里
Rikudō-rin'ne	六道輪廻
Saitō Gesshin	齋藤月岑
Shaku	尺
Shashin	写真
Shihekian Shigetsuta	四壁庵茂蔦
Shinsha	真写
Shoku nihongi	續日本紀
Shōmankō-ji	勝鬘皇寺
Shōrui	生類
Shōsōin	正倉院
Shōtoku taishi denryaku	聖徳太子傳暦
Shōtoku taishi eden	聖徳太子繪傳
Shungyū-zu	駿牛図
Sō	宗
Sō Shiseki	宋紫石
Suiko	推古
Tachibana Morikuni	橘守国
Tani Bunchō	谷文晁
Terajima Ryōan	寺島良安
Tōdai-ji	東大寺
Tokugawa Iemitsu	徳川家光
Tokugawa Ieyasu	徳川家康
Tokugawa Mitsukuni	徳川光圀
Tokugawa Yoshimune	徳川吉宗
Toyamaya Bunzaemon	富山屋文左衛門
Toyotomi Hideyoshi	豐臣秀吉

Tsukuri-mono	作り物
Ueda Kōchō	上田公長
Unkin zuihitsu	雲錦随筆
Utagawa Yoshitoyo	歌川芳豐
Wagō rakuda no sekai	和合駱駝之世界
Wakan sansai zue	和漢三才図会
Wasure nokori	わすれのこり
Zōho bukō nenpyō	增補武江年表
Zoku enseki jisshu	續燕石十種
Zoku Nihon zuihitsu taisei	續日本隨筆大成

Bibliography

Primary Sources

Akatsuki Kanenari 暁鐘成. *Unkin zuihitsu* 雲錦随筆, vol. 4 (1862). In *Nihon zuihitsu taisei* 日本隨筆大成, Nihon zuihitsu taisei henshūbu 日本随筆大成編輯部 ed., series 1, vol. 3. Tokyo: Yoshikawa kōbunkan, 1975.

Jinsaiō 塵哉翁. *Kōgai zeisetsu* 巷街贅説, vol. 2 (1829). In Senzō 森銑三and Kitagawa Hirokuni 北川博邦 eds. *Zoku Nihon zuihitsu taisei* 続日本随筆大成, supplement 9. Tokyo: Yoshikawa kōbunkan, 1983.

Kōnantei Karatachi 江南亭唐立. *Wagō rakuda no sekai* 和合駱駝之世界. Illustrated by Utagawa Kuniyasu 歌川國安. Edo: Moriya Jihei, 1825.

Matsura Seizan 松浦靜山. *Kasshi yawa* 甲子夜話, vols. 8 (1822), 53 (1824). In *Kasshi yawa*, Nakamura Yukihiko 中村幸彦and Nakano Mitsutoshi 中野三敏 eds., vols. 1, 4. Tōyō bunko 東洋文庫 306, 333. Tokyo: Heibonsha, 1977-1978.

Ōtsuki Gentaku 大槻玄澤. "Rakuda mondō 駱駝問答" (1824). In *Ran'en tekihō* 蘭畹摘芳 (1831). Manuscript. Tokyo National Museum.

Saitō Gesshin 齋藤月岑. *Bukō nenpyō* 武江年表 (1850). In Kaneko Mitsuharu 金子光晴 ed. *Zōho bukō nenpyō* 増訂武江年表, vol. 2. Tōyō bunko 118. Tokyo: Heibonsha, 1968.

Shihekian Shigetsuta 四壁庵茂蔦. *Wasure nokori* 忘れ残り, vol. 2 (1824). In *Zoku enseki jisshu* 續燕石十種, vol. 2. Tokyo: Chūō kōronsha, 1980.

Secondary Sources

Asakura Musei朝倉無聲. *Misemono kenkyū* 見世物研究 (1928). Kyoto: Shibunkaku shuppan, 1977.

Chaiklin, Martha. "Exotic-Bird Collecting in Early-Modern Japan," in Gregory M. Pflugfelder and Brett L. Walker eds. *JAPANimals: History and Culture in Japan's Animal Life*. Ann Arbor, MI: Center for Japanese Studies, University of Michigan, 2005, pp. 125-160.

Fuchu Art Museum, ed. *Cute Edo Paintings*. Fuchu: Fuchu Art Museum, 2013.

Fuchu Art Museum 府中市美術館, ed. *Dōbutsu no e: Nihon to Yōroppa, fushigi kawaii heso magari* 動物の繪: 日本とヨーロッパ, ふしぎ・かわいい・へそまがり. Tokyo: Kōndansha, 2021.

Fukuoka, Maki. *The Premise of Fidelity: Science, Visuality, and Representing the Real in Nineteenth-Century Japan*. Stanford: Stanford University Press, 2012.

Higuchi Hiroshi 樋口弘. *Nagasaki ukiyo-e* 長崎浮世絵. Tokyo: Mitō sho'oku, 1971.

Itakura Masaaki板倉聖哲. *Ri Kōrin* Goba zu 李公麟「五馬圖」.Tokyo: Hatori shoten, 2019.

Kano Hiroyuki 狩野博幸. *Kōrin geijutsu no kisō* 光琳藝術の基層. Nihon no bijutsu 日本の美術 462. Tokyo: Shibundō, 2004.

Kawazoe Yū 川添裕. "Hakurai dōbutsu to misemono 舶來動物と見世物," in Nakazawa Katsuaki 中澤克昭 ed. *Rekishi no naka no dōbutsu tachi* 歴史のなかの動物たち, in Hito to dōbutsu no Nihon shi 人と動物の日本史, vol. 2. Tokyo: Yoshikawa kōbunkan, 2009, pp. 127-160.

Laver, Michael. *The Dutch East India Company in Early Modern Japan: Gift Giving and Diplomacy*. London and New York: Bloomsbury Academic, 2020.

Marcon, Federico. "All Creatures Great and Small: Tokugawa Japan and Its Animals," in Robert T. Singer and Kawai Masatomo eds. *The Life of Animals in Japanese Art*. Princeton: Princeton University Press, 2019, pp. 29-37.

Nagoya-shi hakubutsukan 名古屋市博物, ed. *Ehon rakuda gushi* 繪本駱駝具誌. In *Nagoya-shi hakubutsukan shiryō sōsho* 名古屋市博物館資料叢書, vol. 3. Nagoya: Nagoya-shi hakubutsukan, 2007.

Nakano Genzō 中野玄三. *Nihonjin no dōbutsu ga: Kodai kara kindai made no ayumi* 日本人の動物画: 古代から近代までの歩み. Tokyo: Asahi shinbunsha, 1986.

Sugita Hideaki 杉田英明. "Rakuda to Nihon jin: Dōbutsu hyōshō o tōshite mita ikoku shumi 駱駝と日本人: 動物表象を通して見た異國趣味," *Hikaku bungaku kenkyū* 比較文學研究86 (2009), pp. 42–77.

Suntory Museum of Art, ed. *Tani Bunchō: Commemorating the 250th Anniversary of His Birth*. Exhibition catalogue. Tokyo: Suntory Museum of Art, 2013.

Suzuki Hiroyuki 鈴木廣之. "Rakuda o egaku: Maruyama Ōshin hitsu rakuda-zu o megutte ラクダを描く: 圓山応震筆駱駝圖をめぐって," *Bijutsu kenkyū* 美術研究338 (1987), pp. 16-34.

Tobacco and Salt Museum, ed. *Misemono: Spectacles, Shows, and Circuses in the Edo-Meiji Period*. Exhibition catalogue. Tokyo: Tobacco and Salt Museum, 2003.

Tsukamoto Manabu 塚本學. "Mushi o miru me no rekishi: Edo jidai jin o chūshin ni 蟲を見る目の歴史: 江戸時代人を中心に," *Shakaishi kenkyū* 社会史研究6 (1985), pp. 37-87.

Yamane Yūzō 山根有三. *Konishi-ke kyūzō Kōrin kankei shiryō to sono kenkyū: Shiryō* 小西家舊蔵光琳関係資料とその研究: 資料. Tokyo: Chūōkōron bijutsu shuppan, 1962.

Zohar, Ayelet. *The Curious Case of the Camel in Modern Japan: (De)colonialism, Orientalism, and Imagining Asia*. Leiden: Brill, 2022.

About the Author

Hiroyuki Suzuki is Professor Emeritus of Tokyo Gakugei University and Director of Tōyama Memorial Museum. He studies Japanese art history and is presently preparing a book on the history of the formation of "Japanese art history" in the Meiji period. His books include *Antiquarians of Nineteenth-Century Japan. The Archaeology of Things in the Late Tokugawa and Early Meiji Periods* (2022), *Meisho fūzoku-zu* (Genre paintings of famous places) (2007), *Kanō Hideyori hitsu Takao kanpū-zu byōbu. Kioku no katachi* (Maple viewing at Takao by Kanō Hideyori. A shape of memory) (1994), and *Eitoku Tōhaku* (Kanō Eitoku and Hasagawa Tōhaku) (Shinpen meihō Nihon no bijutsu 20, 1991).

6. Pictures of Sea Fish (Haiyu tu) and Knowledge of Nature in Eighteenth-Century China

Ching-Ling Wang

Abstract: This chapter examines an as yet unknown Chinese album, *Pictures of Sea Fish* (*Haiyu tu*) in the collection of the Asian Art Museum in Berlin. It depicts more than 130 species of sea fish. It is a world away from the traditional artistic representation of fish and has more in common with objective scientific investigation of natural history. This article serves as fundamental research of the album. It examines this newly discovered album and subsequently situates it in the different contexts of nature studies and visual culture in early modern China. As the first research on this album, the article sheds light on the role of scientific illustrations within the formation of knowledge of the natural environment in China in the eighteenth century.

Keywords: *Haiyu tu*, Han Liangqing, *Haicuo tu*, Nie Huang, Guangdong

Introduction

In the collection of the Museum of Asian Art, Berlin State Museums, is a heretofore unknown Chinese album, *Pictures of Sea Fish* (Haiyu tu). The original binding of the album has been lost; at some stage it was rebound in Europe as a book. Its cover carries a label in the lower left corner with a collection stamp, a handwritten bibliographic description, and two inventory numbers. The stamp makes it clear that this album was originally in the collection of the library of the Royal Museum of Decorative Arts (now Museum of Decorative Arts, Berlin State Museums). The handwritten description contains three lines:

Fan Lin and Doreen Mueller (eds), *Picturing Animals and Plants in Early Modern China and Japan. Innovation, Experiments, and Anxieties*. Taylor & Francis Group, 2025
DOI: 10.5117/9789048559091_CH06

> Betsu-tschi-tschu-jin [Japanese: Betsuchi Shujin, Chinese: Miechi Zhuren][1]
> Bilder von Fischen mit Erläuterungen [Pictures of fish with explanations]
> Chines. Handzeichen. Peking 1739 [Chinese. Hand-drawing. Beijing 1739]

The first line gives the Japanese pronunciation of the author's Chinese name, Master Miechi (Miechi Zhuren). The second is a German title or description of the album; the third line indicates that it contains Chinese drawings and gives the place (Beijing) and the date (1739) of its creation. There are two inventory numbers: 'J532mtl.' and 'Bi 95, 403'. The letter 'J' indicates that this album was purchased in Japan (which explains the Japanese transcription of the author's Chinese name), the acronym 'Bi' stands for library (Bibliothek), and the number '95' indicates it was purchased or incorporated into the collection in 1895. This album was possibly transferred from the library of the Royal Museum of Decorative Arts to the collection of the Museum of East Asian Art (now Museum of Asian Art) in the 1950s on the occasion of its reopening after the Second World War.[2] But as there is no inventory number for the latter, it is not known when exactly this album was placed in the stacks.

I first found this extraordinary album in storage in 2016 when I was curator of Chinese art at the Museum of Asian Art. The significance of the illustrations, as I will demonstrate, is apparent. Since it is unknown to the field, this article serves as fundamental research of the album. It first examines this newly discovered album, and subsequently situates it in the different contexts of nature studies and visual culture in early modern China. Being the first research on this album, the article sheds light on the role of scientific illustrations within the formation of knowledge of the natural environment in China in the eighteenth century.

The Album, *Pictures of Sea Fish*

The album, *Pictures of Sea Fish*, has eighteen leaves in total (Figures 6.1 and 6.2). The first leaf is the title page with its original Chinese title *Haiyu tu*, written in running script. The second and third leaves contain a preface signed by Master Miechi (Miechi Zhuren). The following fourteen leaves, which are the album's main content, present pictorial representation of more than 130 species of sea fish and other marine creatures, such as squid and octopus. The final leaf has a colophon written by Wang Jian.

1 The Japanese spelling here is unconventional, hence a bit confusing.

2 I thank Prof. Willibald Veit, emeritus director of the Museum für Asiatische Kunst for sharing this information in email correspondence.

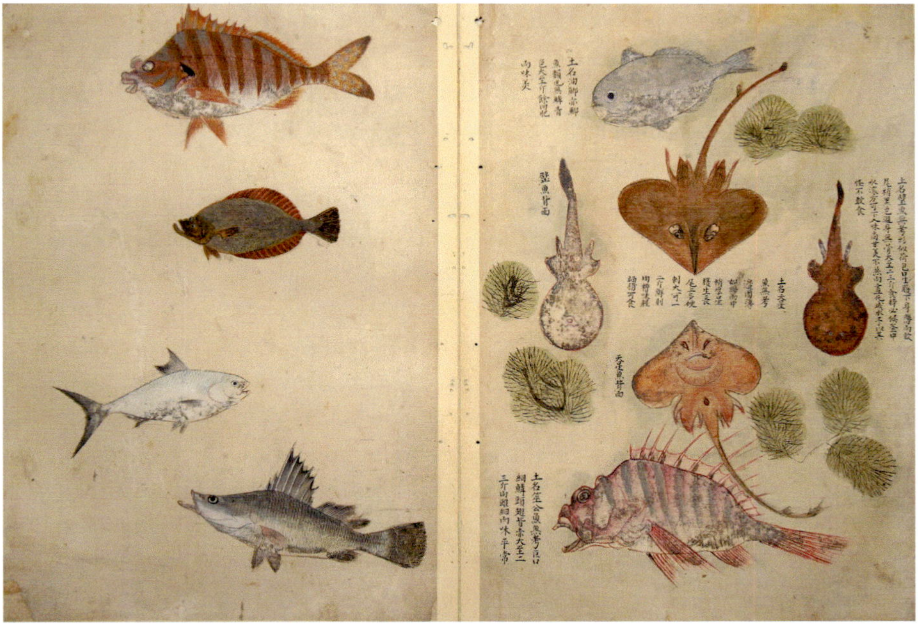

Figure 6.1: Han Liangqing, *Haiyu tu*. 1739. Manuscript, p. 4. Museum für Asiatische Kunst, Staatliche Museen zu Berlin. Photo by Ching-Ling Wang.

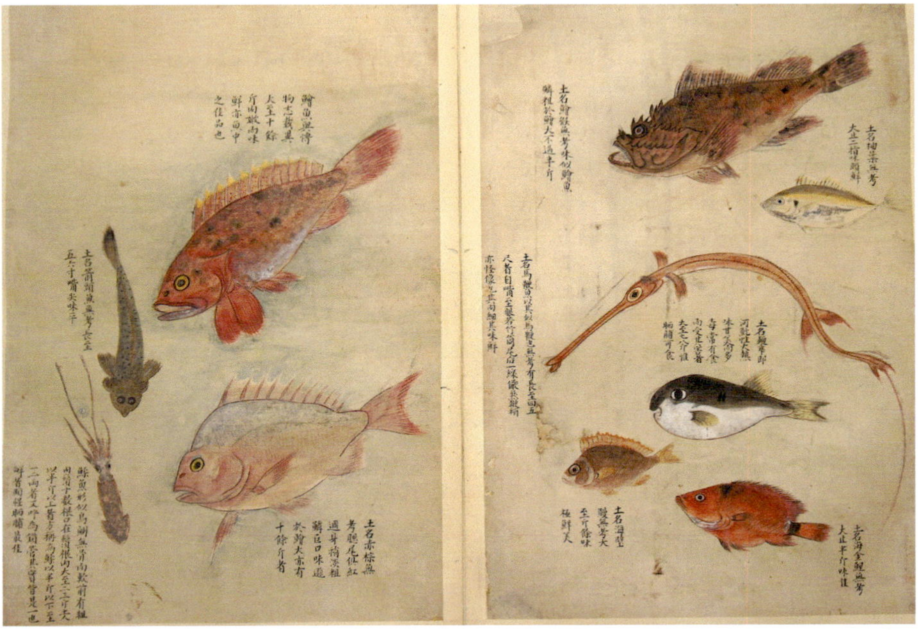

Figure 6.2: Han Liangqing, *Haiyu tu*. 1739, p. 5. Museum für Asiatische Kunst, Staatliche Museen zu Berlin. Photo by Ching-Ling Wang.

The preface signed by Master Miechi explains why and how the album was made:

> Things are normal by nature; [only] those who have limited knowledge and scanty information think they are strange. How can a thing be strange? No one in this world would think that the sun, the moon, a mountain, a river, a horse, cattle, a chicken, and a dog are strange, because they are accustomed to them. I have been posted as an official in the western regions for more than thirty years and have travelled frequently to the front. I never thought that the animals and plants I encountered through my eyes and ears were strange.
> In the *dingsi*-year [1737] I was entrusted with the surveillance of [the coastline] at Jieshi in east Guangdong. When at sea on patrol, [I saw it] as a boundless and vast expanse of water. The waves and the spray were high, I suspected that there were unexpected creatures on the blurry boundary between sky and water. I dived into the water, the servants and sailors were all frightened and did not know what to do, but I bore it with equanimity. What I heard with my ears and saw with my eyes [while at sea], the sounds are like thunder, or like drums and trumpets, or like the sounds of nature under the sky in autumn; the shapes are like pointed peaks and hard rocks, or like architecture, human figures, chariots and horses. They came and went continually and without warning and changed infinitely within a day. But what startled and surprised us was the interaction between winds, clouds, and waves, and creatures: fish, turtles, and shells. 'Odd,' I sighed, 'this is really strange. Why are the things I encounter at sea so unusual?'[3]

As we can see, the text begins with a brief philosophical discourse on what is 'normal/common' and what is 'strange/uncommon.' It further indicates (see below) that its author, Master Miechi, commissioned the album. He had worked in the desert regions of China for more than thirty years before he was put in charge of coastline surveillance at Jieshi, in Guangdong, in 1737. In his new post, he encountered scenery, climate, and creatures that were unusual to him. But who is Master Miechi? Since the text mentions his governmental posts and dates, he can be identified with the help of the *Biographies of officers of the Qing Dynasty* (Guoshi dachen liezhuan); the official posts mentioned by Master Miechi in the preface match the record of a certain Han Liangqing (?–1740):

> [Han] Liangqing held the *jinshi* title for martial arts and was given the position of imperial guard. In the fifty-eighth year of Kangxi's reign [1719] he was appointed lieutenant [*shoubei*] of the central camp at Xining town in Shaanxi province. In the sixtieth year [1721] he was transferred to Lanzhou city as major [*youji*] of the

3 See appendix, no. 1.

> guarding camp. […] In the second year of Qianlong's reign [1737] he was promoted to major general of Jieshi in Guangdong, and later transferred to Shaanxi as major general of Suzhou. In the ninth month of the fifth year [1740] he was given the position of lieutenant general [*tidu*] of Gansu and he died in the eleventh month.[4]

It seems clear that Master Miechi can be no other than Han Liangqing. In the preface, Master Miechi further writes:

> The cook prepared sea fish, the taste was rather delicate. I asked the servant what kind of fish it was. He answered: 'It was what we bought on the market today.' The fish all had strange appearances, some with their fins spread like arms and lifted like a beard, others with radiant colours like makeup. All these different kinds of fish made my heart palpitate. I asked the cook [about their names], he told me they were, 'Just fish.' Then I realised that I had never experienced how strange they actually were. Now that I have been on patrol at sea, I do know. This is why I commissioned the illustrations [of the fish]: to show them to those who will find them strange because they have never seen them. To them [i.e. local people], these [matters are normal and] not worthy of mention.
> I regularly asked fishermen to go out to catch fish, turtles, and shells, had all of these sent to my local magistrate's bureau, and asked a painter to depict them. My intention was to arrange them according to different categories, starting with fish. I collected nearly one hundred and thirty kinds of them. I did not know the names of most of them, so I recorded the names used by local people, and added comments [in the album].[5]

The fish were "normal" and "just fish" for the locals, but for Master Miechi it was another matter. With the help of the local fishermen, Han Liangqing was able to conduct his research on local fish, based on what they had caught. He also involved a local painter to visually document the species. It is worth noting that the local painter hired by Han Liangqing drew the fish in a descriptive and naturalistic way without following established schemes or patterns. The painter focused on capturing the colours of the fish and their biological features, such as the shape of the fins, and the texture and colour of the scales, and rendered them in a lively sketch-like way. For instance, from the depictions of *bi*-fish (*biyu*) and *shigong*-fish (*shigongyu*) on the fourth leaf one can tell that the texture of the body of the former is soft and without

4 *Guoshi dachen liezhuan* (Qing guoshiguan edition), vol. 98, pp. 125–126, archive number: 701005630, collection National Palace Museum, Taipei. My translation, original text see appendix, no. 2; I thank Dr. Zhou Wei-Chiang from the Hong Kong Palace Museum for helping me identify Han Liangqing.

5 See appendix, no. 1.

scales whereas that of the latter is hard and has tiny scales. In the representation of *shigong*-fish the painter paid special attention to the depiction of the shape of the head with its bulging eyes, the spiky and thorn-like dorsal fin and the colour gradations of the scales. He prioritised capturing the specimens' physical features over pursuing a certain quality of line or painterly effect. Since most of the fish are symmetrical, only one side of the fish is shown in the album; in cases of asymmetry, both sides are shown – the *bi*-fish just mentioned is a case in point (fig.1). One might say that the depictions in *Pictures of Sea Fish* show the intention to be faithful to the natural appearance of the fish. The arrangement of fish in the album appears to be random and, on occasion, the painter also paints seagrass in the background with a light colour wash, which gives the viewer the impression that it is swaying in the sea.

The comments accompanying the images of the fish follow a fixed format: first, they document the names by which the fish was known locally; second, they attempt to find a more common name in order to identify the fish; third, they document the physical attributes of the fish including its colour, shape, and size; and fourth, they describe how the fish tastes and occasionally also how local people cook it. For example, in the case of the so-called horsewhip-fish, which in English is known as the red cornetfish (*Fistularia petimba*), Han Liangqing wrote:

> The local name is horsewhip-fish [*mabianyu*], because its shape is like a horsewhip, the actual name is unknown. Its length can be four to five *chi* [1 *chi* = 33.33 cm.], The part from its mouth to its eyes looks like a bamboo joint, the tip of its tail is like a line, just like the tip of a whip. Its appearance is also odd. Its flesh is tender, and it has a delicate flavour.[6] (Figure 6.2)

Han Liangqing tasted most, but not all the fish himself. About the fish he refers to as *bi*-fish he wrote, for example:

> The local name is *bi*-fish [*biyu*], actual name unknown. Its shape resembles a purse [*hebao*]. Its mouth is on the side of its stomach. Its body is thin and soft, the tip of its tail is blackish. The whole body is boneless. It can grow to around two to three *jin* [1 *jin* = 604.8 gr.]. When cooking it, one must wait until the water is boiling, then place it into the water, otherwise the flesh will dissolve in the water. The flavour is somewhat sweet and refreshing. I found it odd, so I did not dare eat it.[7] (Figure 6.1)

He also documented some poisonous fish, for example a type of globefish local people called *mianguai*:

6 See appendix, no. 3.

7 See appendix, no. 4.

> The local name is *mianguai*. Just like the globefish its character [as food] is extremely hot [*re*] in nature; its flavor is sweet and refreshing, but it is poisonous. People are often poisoned because of eating it. It can grow to seven or eight *jin*. It can only be eaten when dried.[8] (Figure 6.2)

Not all fish in the album are accompanied by text. This suggests that the album is unfinished. It would seem that only the fish that have texts were researched by Han Liangqing and that those without texts had yet to be studied or identified. According to his preface, this was intended to be a long-term project; he planned to first collect all local fish, then shells and other entities.

> [...] The number I have now is just ten or twenty percent of all the various species. This project is still incomplete; the shells have not yet been depicted.
> Not so long ago I went up north to present myself before the emperor and received the emperor's order to transfer back to the western regions. This is why the project was never finished. Ah, it was probably the wish of Hairuo [the Sea God] not to have the names of these fish made known to the world! I am also afraid that those who are not used to seeing strange fish like these would think they are just too peculiar and would find this work ridiculous. Although only ten percent has so far been completed, I hope that in the future there will be someone with great knowledge who will collect all species and complete it, to broaden our knowledge.
> Two days before the Chongyang festival [ninth day of the ninth month] in the fourth year [1739] of Emperor Qianlong's [r. 1736–1795] reign, inscribed by Master Miechi [Miechi Zhuren] at the governmental bureau in Jiuquan.[9]

Unfortunately, this project was interrupted when Han Liangqing was posted back to the west regions at Gansu in 1739 and eventually died there in 1740 – *Pictures of Sea Fish* is only the first part of his ambitious project.

Pictures of Sea Fish in the Contexts of Nature Studies and Visual Culture in China

Fish have been depicted in China as early as the Neolithic period. Many archaeological excavations have brought to light images of fish or fish pattern decoration on pottery made in this period. It reflects the lifestyle of a fishing and hunting

8 See appendix, no. 5.
9 See appendix, no. 1.

society.[10] In archaeological finds of later periods and dynasties, such as Zhou (510–314 BCE), Qin (221–207 BCE), and Han (206 BCE–220 CE), fish appear as a common motif in the decoration of objects, such as bronzes, lacquer ware, stone reliefs, tiles, and so on. But although the depiction of fish in China can be traced back to Neolithic pottery, it did not become a major subject for painting until the tenth century.

The earliest textual record relating to fish as a subject of representation in painting history appears in *A Record of the Famous Painters of All the Dynasties* (Lidai minghua ji) written by Zhang Yanyuan (active in the ninth century). Here, Zhang documents an ancient cartographic painting titled *Yellow River Map with Dragons and Fish* (Longyu hetu), and mentions a painter by the name of Xu Miao (171–249) from Wei, one of the Three Kingdoms (220–280), who was proficient at painting fish.[11] According to *A Record of the Famous Paintings of the Tang Dynasty* (Tangchao minghua lu) by Zhu Jingxuan (active in the eighth century), the painter Zheng Qian (691–759) was also a capable painter of fish whose works were praised by his contemporaries.[12]

By the time of the Five Dynasties (907–960) and Northern Song (960–1127), fish painting had become established as a specific genre, as is evident from the *Xuanhe Catalogue of Paintings* (Xuanhe huapu, 1120), the catalogue of the painting collection of Emperor Huizhong (r. 1100–1126), which contains twenty chapters. The recorded paintings were divided into ten categories or genres, with 'dragons and fish' listed as one of them.[13] Famous painters such as Teng Changyou (d. 881), Xu Xi (886–975), Xu Chongsi (active in the tenth century), as well as many others, enjoyed the reputation of being skilled in painting fish. The fact that dragons and fish were seen as belonging to the same category or genre is probably due to a legend describing the transformation of a carp into a dragon or to the Buddhist and Daoist use of images of dragons and fish to pray for rain.[14]

Not only did it become one of the major painting genres, the representation of fish also reached a peak in naturalistic depiction in the Song dynasty (960–1279).

10 See Zhongguo kexueyuan kaogu yanjiuyuan ed., *Xinzhongguo de kaogu shouhuo* (Beijing: Wenxu, 1961), p. 10; Xiaonan Yang ed., *The Golden Age of Chinese Archaeology: Celebrated Discoveries from the People's Republic of China* (Washington D.C.: National Gallery of Art, 1999), pp. 60, 64–67.

11 Zhang Yanyuan, *Lidai minghua ji*, reprint (Beijing: Renmin meishu, 1963), ch. 3, p. 73; ch. 4, pp. 82, 104.

12 Zhu Jingxuan, *Tangchao minghua lu*, reprint in Pan Yungao ed., *Tang Wudai hualun* (Changsha: Hunan meishu, 1997), p. 113.

13 *Huanhe huapu*, 1120, reprint (Taipei: Taiwan shangwu, 1967). For painting activities during the Huizong court and the establishment of different painting genres, see Yun-Ru Chen, "Hua yi yi ye: Chonggu Song Huizongchao de huaihua huodong," Ph.D. dissertation, Graduate Institute of Art History, National Taiwan University, 2008.

14 Tokyo National Museum ed., *Kisshō: Chūgoku bijutsu ni komerareta imi* (Tokyo: Tokyo National Museum, 1998), pp. 240–255.

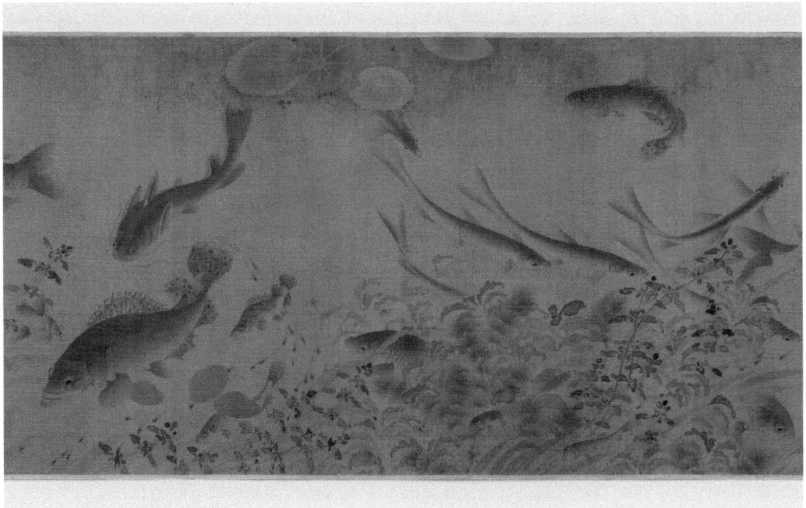

Figure 6.3: Liu Cai, *Luohua youyu tu*. 12th century. Detail. Saint Louis Art Museum.

For example, the scroll *Fish Swimming amid Falling Flowers* (Luohua youyu tu, Figure 6.3), attributed to Liu Cai (fl. 1080–1120) and *Fish and Waterweeds* (Yuzao tu), attributed to Fan Anren (active in the mid-thirteenth century) are considered masterpieces of fish painting.[15] Both are a symphony of rhythm and movement and depict the impression of swimming, darting, drifting fish, and clusters of fish. According to zoologist Dietrich Neumann's detailed observation, in the scroll *Fish and Waterweeds*, Fan Anren depicts a total of forty-seven fish known as sharpbelly (*Hemiculter*), each caught in a different fleeting movement, allowing the viewer to follow and understand the typical sequence of the reproduction process of the fish: from the tracking of sexually mature females (spawners) by males that are ready to mate (milters) up to the males' whirling in circles while releasing their seminal fluid over the eggs laid among aquatic plants by the females. No other painting captures the reproduction process of fish in a similarly logic and vivid manner.[16]

After the Song dynasty, images of fish, both in painting and in the applied arts (such as ceramics, textiles, lacquerware, bronzes and so on), often contained something like a rebus and conveyed auspicious symbolic meanings; for example, goldfish (*jinyu*) would symbolise "gold and jade filling the hall" (*jin yu man tang*); a combination of lotus and fish would symbolise "every year ends with an ample

15 Li Lincan, "Yucao hua de houpo sheenyi," in *Zhongguo meishushi gao* (Taipei: Xiongshi, 1987), pp. 201–206.

16 Dietrich Neumann, "Experiencing and Depicting Nature," in Dietrich Neumann and Josefina Ogando, *Fascinated by Nature: Landscapes, Plants and Animals in the Tradition of Chinese and Japanese Painting from the Neumann-Ogando Collection* (Berlin: Staatliche Museen zu Berlin, 2012), p. 22.

surplus" (*lian nian you yu*), etc.[17] In some rare cases, fish became a vehicle for the artist to express his emotions and philosophy of life, for instance the fish depicted in Bada Shanren's (Zhu Da, 1626–1705) paintings, where the fish are portrayed in a comic and exaggerated manner, almost like a caricature, especially in the eyes, which reflects the contempt and discontent of the painter towards the circumstances of his time.[18]

The illustrations in *Pictures of Sea Fish*, however, do not carry any auspicious symbolic meanings and are not meant to be 'artistic'. They clearly function as illustrations of natural studies and in this respect are closer to illustrations in the tradition of naturalistic enquiry into animals, fungi and plants, and inorganic material, such as minerals, used as *materia medica* (*bencao*) in traditional Chinese medicine.

The Chinese pictorial tradition of illustrated texts on *materia medica* can be traced back to the Sui dynasty (581–619). *A Record of the Famous Painters of All the Dynasties,* mentioned earlier, documents several works, such as *Exemplary Illustrations from Shennong's Canon of Materia Medica'* (Shennong bencao litu), *Exquisitely Executed Materia Medica Illustrations* (Lingxiu bencao tu), and *Illustrations of Materia Medica* (Bencao tu).[19] Although we do find illustrations of fish in books on *materia medica*, for example depictions of sharks (*shayu*) in *Revised Zhenghe Edition of Classified and Practical Basic Materia Medica Based on Historical Classics* (Chongxiu Zhenghe jingshi zhenglei beiyong bencao, Figure 6.4), their purpose differs from that of the images in *Pictures of Sea Fish* – they are there to help the reader identify specimens for medical purposes, whereas the images in *Pictures of Sea Fish* serve as a collection of pictorial documents of unknown creatures.

It should also be mentioned that, usually, new woodblock-printed illustrations were made when existing texts on *materia medica* were re-published. The *Revised Zhenghe Edition of Classified and Practical Basic Materia Medica Based on Historical Classics* was compiled in the Northern Song dynasty by Tang Shenwei (c. 1056–1093),

17 For the development of fish painting in China, see Liu Zhigui, *Zhongguo huihua yuanliu* (Changsha: Hunan meishu, 2003), pp. 418–423. For the symbolism of fish, see Noriko Miyazaki, *Kacho sansuiga o yomitoku: Chugoku kaiga no imi* (Tokyo: Kadokawa Sosho, 2003), pp. 157–171; Hou-mei Song, *Decoded Messages: The Symbolic Language of Chinese Animal Painting* (Cincinnati, OH: Cincinnati Art Museum, 2010), pp. 207–244; Tokyo National Museum ed., *Kisshō*, pp. 26–75; for the rebus play of auspicious meaning in Chinese art, see Tokyo National Museum ed., *Kisshō*; in Chinese painting, see Bai Qianshen, "Image as Word: A Study of Rebus Play in Song Painting (960–1279)," in *Metropolitan Museum Journal* 34 (1999), pp. 57–12.

18 Wang Fangyu, Richard M. Barnhart, Judith G. Smith, eds., *Master of the Lotus Garden: The Life and Art of Bada Shanren* (New Haven, CT: Yale University Press, 1990), pp. 102–104, 128–129, 148–151; Hui-Shu Lee, "The Fish Leaves of the *Anwan Album*: Bada Shanren's Journey to a Landscape of the Past," in *Ars Orientalis*, vol. 20 (1990), pp. 69–85; Hui-Shu Lee, "Bada Shanren's Bird-and-Fish and the Art of Transformation," in *Archives of Asian Art*, vol. 40 (1991), pp. 6–26.

19 See Zhang, *Lidai minghua ji*, vol. 3, pp. 77–78.

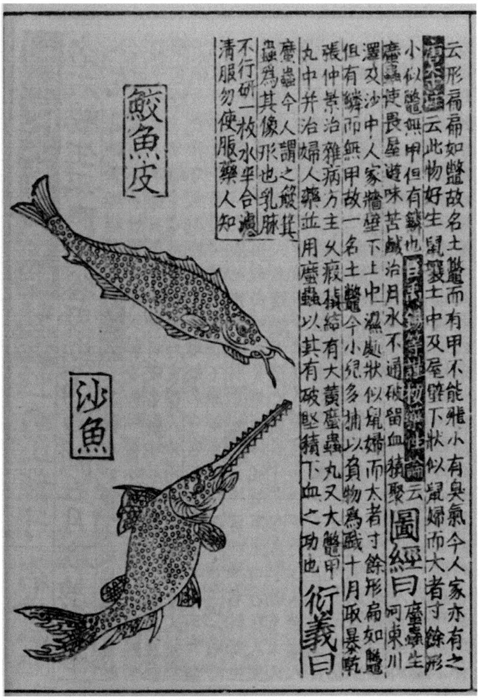

Figure 6.4: Tang Shenwei, *Revised Zhenghe Edition of Classified and Practical Basic Materia Medica Based on Historical Classics* (Chongxiu Zhenghe jingshi zhenglei beiyong bencao) (Beijing: Renmin weisheng, 1957[1249]), p. 434.

a doctor who distilled it from over two hundred Buddhist and Daoist reference works on herbal medicine. The work lists 1,746 herbal medicines; the illustrations were added later when the work was printed in 1249. It was widely known and recommended in medical circles for its rich content and practical applications and continued to be published into the Ming Dynasty (1368–1644). In some cases, the added illustrations and the texts do not match. Also, during the process of transfer from painted illustration to woodblock print and reprint, certain distortions would occur so that the illustrations lack the fidelity of the painted originals, hence, the images in *materia medica* books do not have the same descriptive function as those in *Pictures of Sea Fish*.[20]

In their function as pictorial documents of unknown creatures, the images of fish in *Pictures of Sea Fish* are actually similar to illustrations in books on material culture and nature studies (*pulu*). According to *Complete Library in Four Sections* (Siku quanshu, 1782), *pulu* are a subcategory of traditional Chinese bibliographies

20 Roel Sterckx, "The Limits of Illustration: Animalia and Pharmacopeia from Guo Pu to *Bencaogangmu*," in Vivienne Lo and Penelope Barrett, eds., *Imaging Chinese Medicine* (Leiden: Brill, 2018), pp. 135–150. For illustrations of *materia medica*, see Zheng Jingsheng, "Observational Drawing and Fine Art in Chinese Materia Medica Illustration," in Vivienne Lo and Penelope Barrett eds. (2018), pp. 152–160.

and part of the category of 'Masters and Philosophers' (*zibu*), which was divided into three parts: firstly, books of 'tools and objects' (*qiwu*) including research on antiques, objects associated with the scholar's study, money, machines, and so on; secondly, books of 'eating and drinking' (*yinshan*); and thirdly, books of studies on 'plants and animals' (*caomu qinyu*). The first *pulu* on fish in China is *Illustrated Eulogies of Remarkable Fish* (Yiyu tuzan, preface dated 1544) by Yang Shen (1488–1599). Yang Shen meant his work to be a kind of response to *Illustrations of Exceptional Fish* (Yiyu tu) of the time of the Southern Dynasties (420–589), which, however, has been lost.[21] The book describes eighty-seven kinds of fish and thirty-five kinds of river snails, shells, conches, clams, and other marine life. We might also mention Tu Benjun's (1542–1622) *Notes on the Sea Creatures of Fujian* (Minzhong haicuo shu) published in 1596. The book describes 167 kinds of fish and ninety different types of shells, clams, and turtles. It also includes some freshwater fish, such as carp, and various kinds of frogs and toads. Moreover, the geographical range of the book not only covers the province of Fujian, but also extends to the shores of Guangdong and Zhejiang.[22] Tu quotes from ancient texts and makes use of fishermen's knowledge, to which he adds his own findings. Supplements to both books were published in later periods, but none of them contain any images.[23] The usage and functions of the imagery and the interplay of text and image in *pulu* books are extensively discussed by Martina Siebert in Chapter 7 of this volume.[24]

Fish and their illustrations were also included in encyclopaedias, for example, in the *Illustrated Compendium of the Three Powers* (Sancai tuhui, 1609), intended for common household use (*riyong leishu*) and contained entries on subjects such as astronomy, geography, famous persons, agricultural activities through the year, palaces, tools and implements, the human body, clothing, human affairs, rites and etiquette, jewels, writing, and animals and plants. Another example is the imperial encyclopaedia *Complete Collection of Illustrations and Writings from the Earliest to Current Times* (Gujin tushu jicheng, 1700–1725), a vast work compiled during the reigns of the Qing-dynasty emperors Kangxi (r. 1661–1722) and Yongzheng (r. 1722–1735).[25] The images in *materia medica*, books on material and natural study,

21 See Martina Siebert's article in this volume.

22 Liu Changzhi, "Woguo xian cun zui zao de shuichan dongwu zhi: *Minzhong haicuo shu*," in *Ziran kexue shi yanjiu* 1982: 12, pp. 333–338.

23 For Yang's book, there are both Hu Shi'an's commentary with the title *Yiyu tuzan jian* (Notes on Illustrated Eulogies of Remarkable Fish) and his supplement titled as *Yiyu tuzan bu* (Supplement of Illustrated Eulogies of Remarkable Fish) of the Qing dynasty; as for Tu's book, Xu Bo (1563–1639) wrote a supplement with the title *Minzhong haicuo buzhi* (Supplement of Notes on the Sea Creatures of Fujian).

24 See Chapter 7 in this volume.

25 Chen Menglei and Jiang Tingxi, eds., *Gujin tushu jicheng* (Qing imperial edition, 1700–1725), in the collection of the National Palace Museum, Taipei.

and encyclopaedias were usually added later and are subordinate to the text. The images of fish in *Pictures of Sea Fish*, however, seem to play a more dominant role in the album and they are the source of Han Liangqing's work of natural study and investigation. Moreover, the album uses descriptive images to create an epistemic order.

Han Liangqing's work can also be put into the context of what is known as 'evidential research and learning' (*kaojü* or *kaozheng*), a school and approach of study in late Ming and Qing China from about 1600 to 1850. The approach corresponds to methods of textual studies and is sometimes associated with empirical studies and philology.

One example that is of significance for our context is Nie Huang's (active in the second half of the seventeenth century) album *Pictures of Marine Creatures* (Haicuo tu) of 1698. Nie Huang's work consists of four albums. The total number of illustrated sea creatures and creatures living in the littoral zone in the four albums is three hundred and seventy-one, including fish, crabs, shrimps, turtles, shells, corals, insects, plants, and birds.[26] According to Nie Huang's two prefaces, the reason he started illustrating sea creatures was that it had never been done before. Although there were historical books documenting fish, none of them contained illustrations. The illustrations published in the fish sections included in the books on *materia medica*, according to Nie, were lacking in fidelity.[27] The illustrations appear as afterthoughts or interpretations, rather than observations, and in most cases were inserted later.

Nie Huang's *Pictures of Marine Creatures* is a combination of text and illustration. For each specimen represented in the work, there is an illustration, the creature's name, a poem, and a text describing its appearance, habitual behaviour, and place of habitat. In some cases, Nie Huang even related legends or its application in daily life, such as how to cook it and what it tastes like. Moreover, the text is based on Chinese and European sources. Most of the illustrations are naturalistic depictions of the actual creatures and have been rendered faithfully from life. Some of the illustrations, however, were clearly influenced by images in European works. For example, the image of a whale (which is not actually a fish) in Nie Huang's work (Figure 6.5) was copied directly from the Flemish Jesuit missionary Ferdinandus

26 The first three albums are in the Palace Museum in Beijing and the fourth is now in the collection of the National Palace Museum in Taipei. The complete publication of the first three albums of Nie's *The Marine Creatures* collected in Palace Museum in Beijing, see Gugong bowuyuan ed., *Qinggong haicuo tu* (Beijing: Palace Museum, 2014).

27 Gugong bowuyuan ed., *Qinggong haicuo tu*, pp. 34–37, 40–47. For the fidelity of the images in the natural study books before the use of photography, see Maki Fukuoka, *The Premise of Fidelity: Science, Visuality, and Representing the Real in Nineteen Century Japan* (Stanford, CA: Stanford University Press, 2012); also see footnote 20.

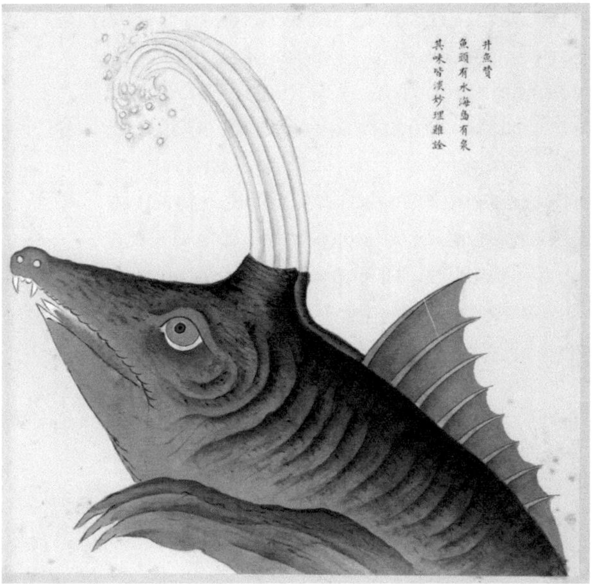

Figure 6.5: Nie Huang, "Whale," in *Pictures of the Marine Creatures* (Haicuo tu), in Palace Museum ed., *Pictures of the Marine Creatures of the Qing Palace* (Qinggong haicuo tu) (Beijing: Place Museum, 2007 [1698]), p. 83.

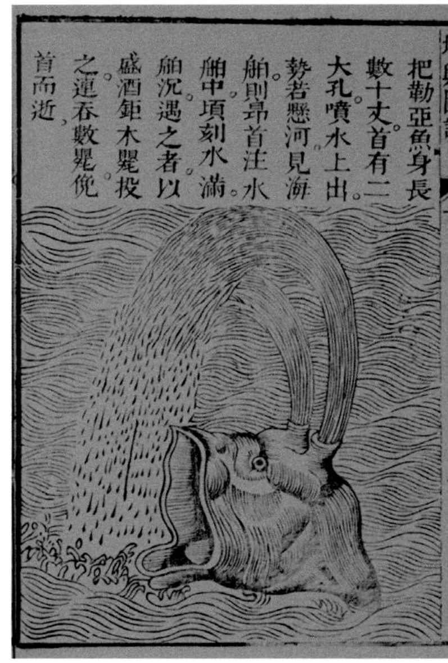

Figure 6.6: "Whale" in Ferdinandus Verbiest, *Illustrated Explanation of the World* (Kunyu tushuo) (1674 edition, Collection of the National Library of France), vol. xia, p. 10.

Verbiest's (also known as Nan Huiren, 1623–1688) book *Illustrated Explanation of the World* (Kunyu tushuo, 1674, Figure 6.6). It is interesting to point out that Nie Huang's *Pictures of Marine Creatures* includes both actual and imaginary creatures, such as the dragon and the merman, which, incidentally, is also the case for the encyclopaedias *Illustrated Compendium of the Three Powers* and *Complete Collection of Illustrations and Writings from the Earliest to Current Times*. The legends he refers to include stories about, for example, shrimps transforming into dragon flies and sharks transforming into tigers. He also includes the legend surrounding the phenomenon of mirages.[28]

Despite the legends and exaggerations, most of the illustrated creatures in Nie's work can be identified as real.[29] According to Nie himself, his method for putting together *Pictures of Marine Creatures* was "first painting the illustration of the specimen then identifying its name, then composing a poem for the species, then conducting textual research on the species, and eventually making the final

28 For the research of *Picture of Marine Creatures*, see Wu Songfeng, "*Haicuo tu*," in *Gugong wenwu yuekan* 363 (2016), pp. 66–73; Zou Zhenghuan, "*Haicuo tu* yu Zhong Xi zhishi zhi jiaoliu," in *Zijincheng* 266 (2017), pp. 124–131.

29 For the identification of the species in Nie's *Pictures of Marine Creatures* in Chinese, English and Latin, see Gugong bowuyuan ed, *Qinggong haicuo tu*, pp. 298–305.

judgement of what the specimen is."[30] Nie also paid attention to the relative size of the species by juxtaposing them, and to the entire visual composition of the album leaf; he arranged the text and the images in a manner full of dynamism and coherence. Nie's work differs from Han Liangqing's in that his *Pictures of Marine Creatures* combined his own investigation with knowledge he had gained from other books (both Chinese and European) and hearsay, whereas Han Liangqing's *Pictures of Sea Fish* was the result of first-hand field investigation.

In 1861, the painter Zhao Zhiqian (1829–1884) produced *Illustrations of Exceptional Fish* (Yiyu tu), a visual record of marine creatures in the regions of Wenzhou and Ruian along the coast of Zhejiang province. Next to each creature he added his commentary (Figure 6.7).[31] His own inscription on the painting states:

> In the *xinyou*-year [1861] of the Xianfeng reign, I [Huishu] travelled to Wenzhou, [I] saw that there are sea creatures with strange appearances, hence I depicted them on this paper and conducted textual research into their names. This is how master painter Gu Kaizhi (345–406) could depict things in a lively manner![32]

Although, on the one hand, the octopus (*zhangjü* [*zhangyu*]), red cornetfish (*mabianyu*), butterfly ray (*yanhong*), and other sea creatures are depicted realistically in this painting, on the other hand, the dolphin (*haixi* [*haitun*]) is depicted as a fish with a pig's head. Furthermore, the so-called dice-fish (*shaiziyu*) depicted in the painting refers to the boxfish (*Ostraciidae*), which oddly looks like a small gaming dice with the head, fins, and tail of a fish: *shaiziyu* literally means 'dice fish', but the name probably merely refers to its boxy shape and spots. Some of the images seem to be based on existing texts rather than direct observation. All of this suggests that Zhao Zhiqian did not see all sea creatures with his own eyes. In this sense, the images in *Pictures of Sea Fish* are more accurate than Zhao's painting. Zhao's painting may be said to fit the context of *kaozheng* study, since his inscription indicates his pursuit of the knowledge of natural history. However, it would seem that he was mainly aiming to depict local species from Wenzhou that he saw as exotic and intriguing. Apart from *Exceptional Fish*, during his stay in Wenzhou, Zhao created several other paintings of plants and fish of the region, for example, four hanging scrolls titled *Plants and Trees in Wenzhou* (Ouzhong caomu tu, in the

30 Gugong bowuyuan ed, *Qinggong haicuo tu*, p. 36; Wu, "*Haicuo tu*," p. 68.

31 Wu Chaoran, "Zhao Zhiqian yibaliuyi nian de sanjian zhuopin *Yiyu tu*, *Ouzhong wuchan tujuan*, *Ouzhong caomu siping*: Jinshi huapai yu Haipai guishu zhi shangque," in Yang Dunyao ed., *Shibian, Xingxiang, liufeng: Zhongguo jindai huihua 1796–1949 xueshu yentaohui lunwenji* (Taipei: Hongxi wenjiao jijinhui, 2008), pp. 451–469; Sheng Wenqiang ed., *Zhao Zhiqian Yiyutu* (Hangzhou: Zhejiang renmin meishu, 2020).

32 For original text, see appendix, no. 6.

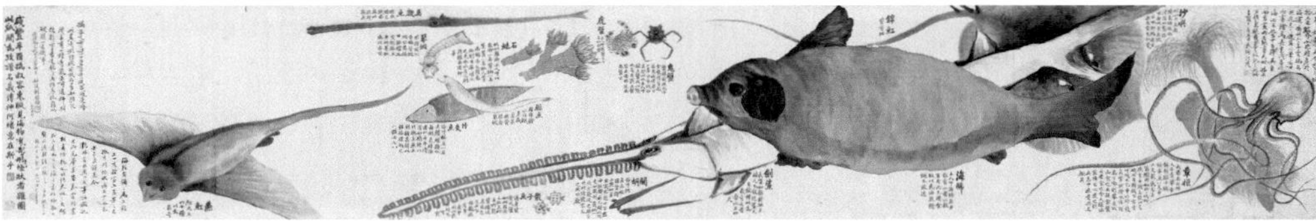

Figure 6.7: Zhao Zhiqian, *Illustrations of Exceptional Fish*. 1861. Wenzhou Yanyuan Museum. Also in Sheng Wenqiang, *The Extraordinary Fish by Zhao Zhiqian* (Zhao Zhiqian Yiyutu) (Hangzhou: Zhejiang renmin meishu, 2020).

collection of Tokyo National Museum) and *Local Products of Wenzhou* (Ouzhong wuzhan tu).[33] In these paintings, the depicted specimens function as natural illustrations and as compositional elements of an artwork at the same time; they are not mere studies of local fish or plants. By depicting these fish and plants in this way, Zhao created a new style within the framework of the traditional category of flora and fauna painting.

Finally, it should be mentioned that the descriptive and naturalistic depiction of fish in *Pictures of Sea Fish* is similar to Western natural history illustrations. The origins of nature studies in Europe can be traced back to Greek and Roman antiquity, but illustrations in the field of natural history studies first flourished during the Renaissance and were seen by contemporary scholars as a 'combination of art and science.'[34] European illustrated books on natural history studies had been introduced to China by missionaries and had their impact on Chinese visual culture. One example showing the impact of Western illustrations is the anonymous and undated album *Manual of Sea Oddities* (Haiguai tu, Figure 6.8), now in the collection of the National Palace Museum, Taipei. Daniel Greenberg first noticed this album and pointed out that its illustrations were based on those in European books on natural science, such as Conrad Gessner's (1516–1565) *Historiae Animalium* (1558), John Jonston's (1603–1675) *Historia Naturalis* (1649–1650) and Johann Zahn's (1641–1707) *Specula Physico-mathematico-historica* (1696), all introduced into China by Jesuit missionaries. Based on the inscription on the cover of the album, Greenberg proposed to date this album to 1688 (*wuchen*-year) and he linked it to the Kangxi emperor, who took a great interest in European sciences.[35] There is, however, no

33 Wan Qingli, *Bingfei shuairuo de bainian: shijiu shiji Zhongguo huihua shi* (Taipei: Xiongshi meishu, 2005), pp. 205–207; Wu, "Zhao Zhiqian yibaliuyi nian de sanjian zhuopin," pp. 451–469; also see the article in this volume by Martina Siebert.

34 Maria Elena De Luca and Gerhard Wolf, "Ligozzis Naturstudien zwischen Kunst und Wissenschaft," in Bundeskunsthalle ed., *Florenz!* (München: Hirmer, 2013), pp. 292–294; For natural history and its imagery in Europe, see Therese O'Malley and Amy R. W. Meyers, eds., *The Art of Natural History: Illustrated Treatises and Botanical Paintings, 1400–1850* (New Haven, CT: Yale University Press, 2008).

35 Daniel Greenberg, "Weird Science: European Origins of the Fantastic Creatures in the Qing Court Painting, the *Manual of Sea Oddities*," in Jerome Silbergeld and Eugene Y. Wang, eds., *The Zoomorphic*

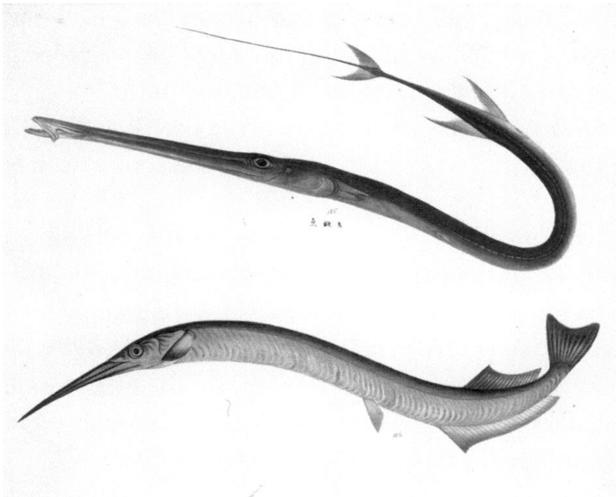

Figure 6.8: Anonymous, *Fistularia petimba, red cornet fish and Strongylura leiura*, ca. 1826-1831, Natural History Museum, London. Also in Judith Magee, *Chinese Art and the Reeves Collection* (London: Natural History Museum, 2011), p. 91.

strong evidence to support this date, as the album could have been painted in later *wuchen*-years, such as 1748, or even 1808 or 1868. It is worth mentioning that *Manual of Sea Oddities* contains no text but only images; its function and purpose are therefore unclear. It may simply have been produced to satisfy curiosity or to provide visual amusement. Furthermore, the style of the depictions resembles export paintings produced in the late eighteenth and nineteenth centuries in Canton. Since the album's production context is unclear, there is also a possibility that it was not produced by the imperial workshop, but made in Canton and presented to the court as a local tribute.[36]

What could have been the context in Canton? Botanical and zoological illustrations were part of the vast scope of export paintings produced in Canton during the second half of the eighteenth century and the nineteenth century. Foreign scholars, such as the botanist John Bradby Blake (1745–1773), the naturalist John Reeves (1774–1856), and others, were engaged in their production.[37] This applies especially to Reeves who was in Canton and personally involved in the making of botanical and zoological illustrations with local painters. During his stay in Canton, he also commissioned four sets of illustrations of fish. According to his

Imagination in Chinese Art and Culture (Honolulu: University of Hawaii Press, 2016), pp. 379–400.

36 In her insightful article, Lai Yu-chih reminds us of the interaction between the court painting and the export painting produced in Canton, see Lai Yu-Chih, "Qinggong yu Guangdong waixiao huafeng de jiaohui: wumingkuan *Haidong cejing tuce* chutan," in *Gugong wenwu yukan* 363 (2013), pp.74–86.

37 Chen Yin, *Lingnan huaniaohua liubian 1368–1949* (Shanghai: Shanghai guji, 2004), pp.265–306; Jiang Yinghe, *Qingdai yanghua yu Guangzhou tongshang kuoan* (Beijing: Zhonghua shujü, 2007), pp. 222–226; Judith Magee, *Images of Nature: Chinese Art and the Reeves Collection* (London: Natural History Museum, 2011), pp. 4–13.

notes, these illustrations were painted by four painters: Akut, Akam, Akew, and Asung.[38] The illustrations commissioned by Reeves painted by Cantonese local painters can be seen as a combination of Chinese art (executed in the Western manner) and European natural history. The artists who worked with Reeves used gold and silver powder to reproduce the iridescence of fish scales. A comparison of the illustration of the red cornetfish in *Pictures of Sea Fish* (Figure 6.2) with the one in Reeves' collection (Figure 6.8) shows that, although the image in Han Liangqing's *Pictures of Sea Fish* is less vivid, it captures the biological characteristics of the fish accurately. Judging from the techniques, however, the local painter commissioned by Han Liangqing may not have been professionally trained and was obviously less skilled than the professional painters active in Canton. On the other hand, due to the enormous volumes of production by various workshops, the quality and accuracy of Canton export paintings also vary.

Despite their 'scientific' aura, Canton illustrations also include imaginary creatures. For example, the collection of the Wereldmuseum Leiden contains twelve fish albums formerly in the possession of Jean Theodore Royer (1737–1807), an amateur sinologist and lawyer from The Hague. The albums may be dated to the late eighteenth century, around 1770. Each contains twenty-four pictures, 288 different fish in all.[39]

In Royer's albums, the fish are depicted individually in a natural setting with background on each leaf and each fish has its Chinese name written next to the illustration. The images have been executed in a Western manner with shading and colour wash to depict the creatures' spots and patterns in detail and to create a three-dimensional representation. Although each album consists of twenty-four leaves, there seems to be no system or order; each album presents a mixture of different species, both freshwater and sea fish. Take the first album for example: it includes a crab, a lobster, various carps, goldfish, several kinds of perches, different squids, loaches, a marble goby (*sunkeyu*, *Oxyeleotris marmorata*), a rhino-fish (*xiniuyu*), and others. Apart from this, most of the names of the fish documented in the albums are not formal names but local names or made-up names, making them difficult to identify.

Although the fish in Royer's albums are portrayed in a naturalistic manner, some of the species do not exist. For example, the *pi*-fish (*piyu*) on leaf nine is a fish with a bird's head, documented in the *Classic of Mountains and Seas* (Shanhaijing), an ancient compilation of mythical geography and beasts.[40] The so-called rhino-fish

38 Judith Magee, *Images of Nature*, p. 11; Kate Bailey, *John Reeves: Pioneering Collector of Chinese Plants and Botanical Art* (London: ACC Art Books, 2019), pp. 108–109.

39 For an overview of Royer's collection, see Jan van Campen, *De Haagse jurist Jean Theodore Royer (1737–1807) en zijn verzameling Chinese voorwerpen* (Hilversum: Verloren, 2000).

40 Inv-no.: RV-360–379a9, Wereldmuseum Leiden. For the animals, plants and fish documented in the *Shanhaijing*, see Zhang Yan, *Shanhaijing yu gudai shehui* (Beijing: Xinhua, 1999), pp. 36–98.

on leaf twenty-three is depicted as a fish with a buffalo's head (and two forelegs). It resembles the buffalo-fish (*qianniuyu*) depicted in Nie's album; the depiction has its source in a legend told by local people.[41] The images in Royer's fish albums come from various sources. Some of the species were probably painted from life, but many of the depictions were based on descriptions in ancient texts or legends and some of them are plainly imaginary, such as the horse-fish (*mayu*) and the monkey-fish (*houyu*).[42]

This kind of export painting produced by anonymous workshop painters in Canton often served as souvenirs – exotic visual sources of wonder or amusement for Western consumers. When examining the content of all twelve albums, one does not find any order or system. Some of the fish appear several times. These illustrations were probably made as single sheets with multiple copies and randomly assembled by the workshop. The cases of John Reeves and Jean Theodore Royer illustrate the different qualities of export painting produced in Canton as well as the different degrees of involvement of Western naturalists.

As in *Pictures of Sea Fish*, the fish depicted in the illustrations commissioned by Reeves were collected at the local market.[43] This suggests that all the fish depicted in Reeves' illustrations are edible. By contrast, Han Liangqing asked the fishermen to bring in everything they caught, including poisonous sea creatures or species unsuitable for consumption. The fish illustrations commissioned by Reeves were sent to England, where they met with the approval of the famous naturalist John Richardson (1787–1865). In his "Report on the Ichthyology of the Seas of China and Japan" of 1845 Richardson wrote:

> John Reeves, [...] who was long resident at Macao, filling an important office in the employ of the India Company, with an enlightened munificence, caused beautiful coloured drawings, mostly of the natural size, to be made of no fewer than 310 species of fish which are brought to the market at Canton. These drawings are executed with correctness and finish which will be sought after in vain in the older works on ichthyology, and which are not surpassed in the plates of any large European work of the present day. The unrivalled brilliancy and effect of the colouring, and correctness of profile, render them excellent portraits of the fish [...][44]

The naturalistic illustrations produced in Canton in collaboration with foreign scholars were sent back to Europe and made an impact on scholarship there. In

41 Gugong bowuyuan ed, *Qinggong haicuo tu*, p. 178.

42 Inv-no.: RV-360-379d22, RV-360-379e22, Wereldmuseum Leiden.

43 Bailey, *John Reeves*, p.108.

44 J.M. Richardson, "Report on the Ichthyology of the Seas of China and Japan," in *Report of the British Association for the Advancement of Science* (1845), quoted from Bailey, *John Reeves*, pp. 113–115.

contrast, these illustrations seem to have had little or no impact in China, and the same holds true for Han Liangqing's *Pictures of Sea Fish*.

Conclusion

Although *Pictures of Sea Fish* was the result of Han Liangqing's own curiosity about sea creatures he had never seen before, the painter he commissioned faithfully documented the fish. The album is Han Liangqing's first-hand field investigation of local fish in Jieshi, Guangdong province. The comments were based on his own observations. He carefully described the appearance of the fish including their size, colour, shape, and features in his text; the taste of the fish he included in the album was likewise based on his direct experience. The text is based on facts and the illustrations are naturalistic.

In addition, Han Liangqing set out to make a complete record and collect as many fish as possible, instead of selecting a few for their aesthetic qualities – he did not have a work of art in mind. Most importantly, the purpose of the album was to inform viewers that these creatures were not strange, but quite normal. This was a world removed from Chinese traditional artistic representation of fish and it foreshadows illustrations of natural history produced later in Canton.

Han Liangqing's method consisted of first commissioning the fish from the local fishermen, then obtaining illustrations from local anonymous painters and subsequently using these for his studies. The illustrations in *Pictures of Sea Fish* thus became his primary source material. This approach differs from the traditional Chinese *kaozheng* method in which text forms the basis of research.

His album *Pictures of Sea Fish* is an extraordinary example of the pursuit of empirical science in eighteenth-century China. The album was made to form part of a book of ichthyological studies. Unfortunately, this original grand project was never accomplished due to Han Liangqing's transfer and death. What we are left with is a remarkable collection of images highlighting an approach which has remained hidden until now.

Appendices

1.

物本常也，而寡聞渺見者以為恠。夫物亦安有怪哉？世固未有以日月山川、馬牛雞犬為怪者，習見之常也。余宦跡西域垂三十年，數四行間頻游塞上，凡耳目所經禽獸草木，未嘗見所為怪者。及丁巳歲，奉命守粵東之碣石，巡行海上，浩渺無垠，波撼氣蒸，天水混茫中疑有靈怪不測者。潛遊其下，僕從、篙師，戰兢股栗，莫之所謂，余安之若素也。耳之所得，目之所遇，或聲如雷霆、如鼓吹、如秋空天籟；或狀如兀峰磐石、如樓臺人物車馬。傾刻之間，往來莫定，一日之內，變幻無窮。蓋無非風雲波濤，魚鱉鱗介之屬相與動心而駭目也。乃喟然曰：「異哉。此真怪矣」。何聞見之奇如此耶？已而。庖人以海魚進食，甚鮮。問其僕:「何魚也」。對曰：「日之所市」。怪狀奇形，張手怒鬣，傅粉施朱，金碧燦爛，種種色色，令人心悸。詢於庖人，皆曰魚也。乃耳目之所未接者，予笑曰而怪乎，向游海上，已知之矣。方將圖以示天下之不見怪而恠者，此何足云。於是數命罟師，摻舟入海，凡日之所得魚鱉鱗介，悉□（破損，缺字）于署，令善畫者逐照而寫之。意以鱗介分類成帙，先自魚始。甫得一百三十餘種，名多無考，因土人之名而註之，不過得種類十分之一二。事未竣，其介族尚未搦管。予適入覲北上，旋奉調西來，遂不果其事。嗟乎!望海若之靈，不欲留諸天壤耶！抑恐不習見者之以為怪而故靳其工耶！姑存什一，俟閎博君子積類而成，以廣聞見之異云。時乾隆四年重陽前二日，蔑癡主人識於酒泉官署。

2.
良卿由武進士授侍衛。康熙五十八年授陝西西寧鎮中營守備。六十年遷蘭州城守營游擊。雍正元年遷莊浪營參將。時青海羅卜藏丹津叛莊浪，賊番助逆，良卿隨涼州總兵楊盡信等擊賊於碁子山有功，賜孔雀翎並銀千兩。四年遷寧夏中尉副將。乾隆二年擢廣西 [字誤，應為：東] 碣石總兵，尋調陝西肅州總兵。五年九月授甘肅提督，十一月卒。

3.
土名馬鞭魚，以其似馬鞭也，無考。有長至四五尺者，自嘴至眼若竹筒，尾後一線，儼然鞭稍，亦怪像也。其肉細，其味鮮。

4.
土名壁魚，無考。形似荷包，口生腹下，身薄而軟，尾稍黑色，週身無骨，大至二、三斤。食時必候釜中水滾，方可下入，味尚甘美，不然肉盡化成水，予以其怪，不敢食。

5.
土名麵乖，即河鮑。性大熱，味甘美而多毒，常有食而受其害者，大至七八斤，惟晒脯可食。

6.
咸豐辛酉，撝叔客東甌，見海物有奇形怪狀者，雜圖此紙，間為考證名義，傳神阿堵，意在斯乎？

Glossary

Bada Shanren (Zhu Da)	八大山人（朱耷）
bencao	本草
Bencao tu	本草圖
biyu	壁魚
caomu qinyu	草木禽魚
chi	尺
Chongyang	重陽
Chongxiu Zhenghe jingshi zhenglei beiyong bencao	重修政和經史證類備用本草
Fan Anren	范安仁
Gu Kaizhi	顧愷之
Gansu	甘肅
Guangdong	廣東
Gujin tushu jicheng	古今圖書集成
Guoshi dachen liezhuan	國史大臣列傳
Haicuo tu	海錯圖
Haiguai tu	海怪圖
Hairuo	海若
Haiyu tu	海魚圖
haixi (*haitun*)	海稀（海豚）
Han Liangqing	韓良卿
Huizong	徽宗
hebao	荷包
houyu	猴魚
Jieshi	碣石
jin	斤
jinyu	金魚
jin yu man tang	金玉滿堂
Kangxi	康熙
Kunyu tushuo	坤輿圖說
lian nian you yu	連年有餘
Lidai minghua ji	歷代名畫記
Lingxiu bencao tu	靈修本草圖
Liu Cai	劉寀
Longyu hetu	龍魚河圖
Luohua youyu tu	落花游魚圖
mabianyu	馬鞭魚
mayu	馬魚

mianguai	麵乖
Miechi Zhuren	篾癡主人
Minzhong haicuo shu	閩中海錯疏
Nan Huiren (Ferdinandus Verbiest)	南懷仁
Nie Huang	聶璜
Ouzhong caomu tu	甌中草木圖
Ouzhong wuzhan tu	甌中物產圖
piyu	魮魚
pulu	譜錄
Qianlong	乾隆
qianniuyu	潛牛魚
qiwu	器物
Wang Jian	王建
Wenzhou	溫州
re	熱
riyong leishu	日用類書
Sancai tuhui	三才圖會
Shanhaijing	山海經
shaiziyu	骰子魚
shayu	鯊魚
Shennong bencao litu	神農本草例圖
Shigongyu	筮公魚
Siku quanshu	四庫全書
sunkeyu	筍殼魚
Tangchao minghua lu	唐朝名畫錄
Tu Benjun	屠本畯
wuchen	戊辰
xingshu	行書
Xu Chongsi	徐崇嗣
Xu Miao	徐邈
Xuanhe huapu	宣和畫譜
Yang Shen	楊慎
yanhong	燕魟
yinshan	飲膳
Yiyu tu	異魚圖
Yiyu tuzan	異魚圖贊
Yongzheng	雍正
Yuzao tu	魚藻圖
zhangjü (*zhangyu*)	章拒（章魚）
Zhang Yanyuan	張彥遠

Zhao Zhiqian	趙之謙
Zheng Qian	鄭虔
Zhu Jingxuan	朱景玄
zibu	子部

Bibliography

Primary Sources

Chen, Menglei 陳夢雷, Jiang, Tingxi 蔣廷錫eds. *Gujin tushu jicheng* 古今圖書集成. 1700–1725, Qing imperial edition, in the collection of the National Palace Museum, Taipei.

Guoshi dachen liezhuan 國史大臣列傳. Qing guoshiguan edition, in the collection of the National Palace Museum, Taipei. vol. 98.

Zhang Yanyuan 張彥遠. *Lidai minghua ji* 歷代名畫記. Reprint. Beijing: Renmin meishu, 1963.

Zhao Ji 趙佶, comp. *Huanhe huapu* 宣和畫譜. 1120, reprint. Taipei: Taiwan shangwu, 1967.

Zhu Jingxuan 朱景玄. *Tangchao minghua lu* 唐朝名畫錄. Reprint in Pan Yungao 潘運告 ed., *Tang Wudai hualun* 唐五代畫論. Changsha: Hunan meishu, 1997.

Secondary Sources

Bai Qianshen. “Image as Word: A Study of Rebus Play in Song Painting (960–1279),” *Metropolitan Museum Journal* 34 (1999): 57–12.

Bailey, Kate. *John Reeves: Pioneering Collector of Chinese Plants and Botanical Art*. London: ACC Art Books, 2019.

Chen Yin 陳瀅. *Lingnan huaniaohua liubian 1368–1949* 嶺南花鳥畫流變 1368–1949. Shanghai: Shanghai guji, 2004.

Chen Yun-Ru 陳韻如. “Hua yi yi ye: Chonggu Song Huizongchao de huaihua huodong 畫亦藝也：重估宋徽宗的繪畫活動.” Ph.D. dissertation, Graduate Institute of Art History, National Taiwan University, 2008.

De Luca, Maria Elena and Gerhard Wolf. “Ligozzis Naturstudien zwischen Kunst und Wissenschaft,” in Bundeskunsthalle ed., *Florenz!*. München: Hirmer, 2013, pp. 292–294.

Fukuoka, Maki. *The Premise of Fidelity: Science, Visuality, and Representing the Real in Nineteenth-Century Japani*. Stanford, CA: Stanford University Press, 2012.

Greenberg, Daniel. “Weird Science: European Origins of the Fantastic Creatures in the Qing Court Painting, the *Manual of Sea Oddities*,” in Jerome Silbergeld and Eugene Y. Wang, eds., *The Zoomorphic Imagination in Chinese Art and Culture*. Honolulu: University of Hawaii Press, 2016, pp. 379–400.

Gugong bowuyuan 故宮博物院 ed. *Qinggong haicuo tu* 清宮海錯圖. Beijing: Palace Museum, 2014.

Jiang Yinghe 江瀅河. *Qingdai yanghua yu Guangzhou tongshang kuoan* 清代洋畫與廣州口岸. Beijing: Zhonghua shujü, 2007.

Lai, Yu-Chih 賴毓芝. "Qinggong yu Guangdong waixiao huafeng de jiaohui: wumingkuan *Haidong cejing tuce* chutan 清宮與廣東外銷畫風的交會ー無名款海東測景圖測初探," *Gugong wenwu yukan* 故宮文物月刊363 (2013): 74–86.

Lee, Hui-Shu. "The Fish Leaves of the *Anwan Album*: Bada Shanren's Journey to a Landscape of the Past," *Ars Orientalis* 20 (1990): 69–85.

Lee, Hui-Shu. "Bada Shanren's Bird-and-Fish and the Art of Transformation," *Archives of Asian Art* 40 (1991): 6–26.

Li Lincan 李霖燦. "Yucao hua de houpo sheenyi 魚藻畫的活潑生意," in *Zhongguo meishushi gao*中國美術史稿. Taipei: Xiongshi, 1987, pp. 201–206.

Liu Changzhi 劉昌芝. "Woguo xian cun zui zao de shuichan dongwu zhi: *Minzhong haicuo shu* 我國現存最早的水產動物志ー《閩中海錯疏》," in *Ziran kexue shi yanjiu* 自然科學史研究1982 (12): 333–338.

Liu Zhigui 劉治貴. *Zhongguo huihua yuanliu* 中國繪畫源流. Changsha: Hunan meishu, 2003.

Magee, Judith. *Images of Nature: Chinese Art and the Reeves Collection*. London: Natural History Museum, 2011.

Miyazaki, Noriko 宮崎法子. *Kacho sansuiga o yomitoku: Chugoku kaiga no imi* 花鳥・山水画を読み解くー中国絵画の意味. Tokyo: Kadokawa Sosho, 2003.

Neumann, Dietrich. "Experiencing and Depicting Nature," in Dietrich Neumann and Josefina Ogando eds., *Fascinated by Nature: Landscapes, Plants and Animals in the Tradition of Chinese and Japanese Painting from the Neumann-Ogando Collection*. Berlin: Staatliche Museen zu Berlin, 2012, pp. 14–30.

O'Malley, Therese and Amy R. W. Meyers, eds. *The Art of Natural History: Illustrated Treatises and Botanical Paintings, 1400–1850*. New Haven, CT: Yale University Press, 2008.

Tokyo National Museum ed. *Kisshō: Chūgoku bijutsu ni komerareta imi* 吉祥ー中国美術にこめられた意味. Tokyo: Tokyo National Museum, 1998.

Song, Hou-mei. *Decoded Messages: The Symbolic Language of Chinese Animal Painting*. Cincinnati, OH: Cincinnati Art Museum, 2010.

Sheng Wenqiang 盛文強ed. *Zhao Zhiqian Yiyutu* 趙之謙異魚圖. Hangzhou: Zhejiang renmin meishu, 2020.

Sterckx, Roel. "The Limits of Illustration: *Animallia* and Pharmacopeia from Guo Pu to *Bencaogangmu*," in Vivienne Lo and Penelope Barrett eds., *Imaging Chinese Medicine*. Leiden: Brill, 2018, pp. 135–150.

Van Campen, Jan. *De Haagse jurist Jean Theodore Royer (1737–1807) en zijn verzameling Chinese voorwerpen*. Hilversum: Verloren, 2000.

Wan Qingli 萬青力. *Bingfei shuairuo de bainian: shijiu shiji Zhongguo huihua shi* 並非衰弱的百年：十九世紀中國繪畫史. Taipei: Xiongshi meishu, 2005.

Wang, Fangyu, Richard M. Barnhart, Judith G. Smith, eds. *Master of the Lotus Garden: The Life and Art of Bada Shanren*. New Haven, CT: Yale University Press, 1990.

Wu Chaoran吳超然. "Zhao Zhiqian yibaliuyi nian de sanjian zhuopin Yiyu tu, Ouzhong wuchan tujuan, Ouzhong caomu siping: Jinshi huapai yu Haipai guishu zhi shangque 趙之謙一八六一年的三件作品《異魚圖》、《甌中物產圖卷》、《甌中草木四屏》：金石畫派與海派歸屬之商榷," in Yang Dunyao 楊敦堯ed., *Shibian, Xingxiang, liufeng: Zhongguo jindai huihua 1796–1949 xueshu yentaohui lunwenji* 世變・形象・流風：中國近代繪畫1796–1949學術研討會論文集. Taipei: Hongxi wenjiao jijinhui, 2008, pp. 451–469.

Wu Songfeng 吳誦芬. "*Haicuo tu* 海錯圖," *Gugong wenwu yuekan* 故宮文物月刊363 (2016): 66–73.

Yang, Xiaonan ed. *The Golden Age of Chinese Archaeology: Celebrated Discoveries from the People's Republic of China*. Washington D.C.: National Gallery of Art, 1999.

Zhang Yan 張岩. *Shanhaijing yu gudai shehui* 山海經與古代社會. Beijing: Xinhua, 1999.

Zheng Jingsheng. "Observational Drawing and Fine Art in Chinese Materia Medica Illustration," in Vivienne Lo and Penelope Barrett, eds., *Imaging Chinese Medicine*, pp. 152–160.

Zhongguo kexueyuan kaogu yanjiuyuan 中國科學院考古研究院ed. *Xinzhongguo de kaogu shouhuo* 新中國的考古收穫. Beijing: Wenxu, 1961.

Zou Zhenghuan 鄒振環. "*Haicuo tu* yu Zhong Xi zhishi zhi jiaoliu《海錯圖》與中西知識之交流," *Zijincheng* 紫禁城 266 (2017): 124–131.

About the Author

Ching-Ling Wang is curator of Chinese art at the Rijksmuseum, Amsterdam. His research interests focus mainly on Chinese literati painting, Ming and Qing court art, and questions concerning visual, material, and cultural exchanges between China and Europe in the seventeenth and eighteenth centuries.

7. Treatise (*pu*) versus Illustration (*tu*): The Absence and Presence of Illustrations in *Pulu* Writings on Chinese Nature Studies

Martina Siebert

Abstract: This chapter examines the complex relationship between text and image in *pulu* writing about animals and plants. *Pulu* began to thrive as a genre in the twelfth century, a time when bibliographer and historian Zheng Qiao coined the notion of image (*tu*) and descriptive text (*pu*) as mutually supportive means for the packaging of knowledge. *Pulu* are categorised on the basis of their topics and, against this background, are often expected to be illustrated, although this is rarely the case. The chapter sets out to explore examples of non-illustrated and illustrated *pulu* and the circumstances and scholarly attitudes that led to their production. The chapter begins with brief probes into the history of the bibliographical classification of *pu*- and *tu*-type books and some reflections on the titles of works.

Keywords: nature studies, *pulu* 譜錄, *tupu* 圖譜

Introduction

The Chinese historical genre of 'treatises and lists' (*pulu*) developed as one specialist mode for scholars to write and read about nature studies and material culture. *Pulu* are defined, on the one hand, by their topic, namely, a type or a group of concrete physical 'things' such as a certain type of plant, animal, or artefact, and, on the other hand, by their format, namely, a structure consisting of a specific set of chapters and/or a list of variants or individuals of the respective 'thing.' Becoming popular in the twelfth century, they continued to thrive until the end of imperial China and

Fan Lin and Doreen Mueller (eds), *Picturing Animals and Plants in Early Modern China and Japan. Innovation, Experiments, and Anxieties*. Taylor & Francis Group, 2025
DOI: 10.5117/9789048559091_CH07

beyond.[1] As they describe 'things,' it seems almost a matter of fact that the texts would be accompanied by illustrations. This expectation was probably reinforced by the notion of *tupu*, i.e., the combination of illustration (*tu*) and treatise (*pu*), as the most reliable means to transmit knowledge, propagated by the Song-dynasty historian and philosopher Zheng Qiao (1104–1162). Historians of China have thus often been misled to assume that *pulu*, which began to thrive at around the same time, would adhere to that model.

This chapter aims to uncover the complex relation between *pu* (treatise/text) and *tu* (illustration/drawing) within *pulu* writing. It will consider the various types of illustrations in *pulu* writing and the significance of their presence or absence to examine the modes used by *pulu* authors to package, valorise, and redistribute knowledge. *Pulu* as a genre served Chinese authors[2] during the *longue durée* from the eleventh to the early twentieth century to present and frame specialist knowledge on animal and plant species without addressing overarching (Western) concepts such as botany or zoology. After the introduction of Western science, authors employed modernised versions of *pulu* to serve their audience of aficionado communities such as orchid cultivators and goldfish breeders among others. However, they retained genre-specific modes of presenting their knowledge and sometimes even developed their own 'modern' forms of illustration. Other groups of *pulu* readership, namely, those reading *pulu* for erudite enjoyment and those reading about a specific *pulu* topic to acquire an overview of all there was to know about a specific 'thing,' partially moved on to other genres. Nevertheless, in any modern book on, for example, chrysanthemums one often still finds references to historical *pulu* writings and their expertise.

Before embarking on discussing examples of *pulu* titles and their use or non-use of illustrations, the first section of this chapter addresses two more general questions.

1 For a short introduction on the development of *pulu* as a genre and bibliographical category in historical China see Martina Siebert, "Neue Formen für neue Themen – *pulu* als bibliographische Kategorie und als Schriften zu Sach- und Naturkunde," in Florian C. Reiter, ed., *Das Reich der Mitte – in Mitte. Studien Berliner Sinologen* (Wiesbaden: Harrassowitz, 2006), pp. 59–70; for a more in-depth analysis, see Siebert, *Pulu – 'Abhandlungen und Auflistungen' zu materieller Kultur und Naturkunde im traditionellen China* (Wiesbaden: Harrassowitz, 2006). On the notion of possessing the objects described in *pulu* or constructing the knowledge about them as referenceable truth within the scholarly world of texts, see Siebert, "Consuming and Possessing Things on Paper: Examples from Late Imperial China's Natural Studies," in Elif Akçetin and Suraiya Faroqhi, eds., *Living the Good Life: Consumption in the Qing and Ottoman Empires of the Eighteenth* (Leiden and Boston, MA, Brill 2017), pp. 384–408; and Siebert, "Animals as Text: Producing and Consuming 'Text-Animals'," in Siebert, Roel Sterckx, and D. Schäfer, eds., *Animals Through Chinese History* (Cambridge: Cambridge University Press, 2018), pp. 139–159.

2 Many *pulu* works did rely heavily on the collection and arrangement of existing text snippets. I will nevertheless call them 'authors' instead of 'compilers,' etc. as their role in the titles themselves and in historical bibliographies rarely is labelled as 'compiled by' but most often as 'written by.'

It firstly examines how *tu* (illustration) played out in the book titles of *pulu* and secondly follows some traces in early Chinese bibliographical classification that suggest a special perception and classificatory treatment of illustrated works and works of a treatise or register type (*pu*). The second section of this chapter reflects on various aspects relating to non-illustrated *pulu* using the popular *pulu* topic of chrysanthemums as an example. The third section introduces works in which paintings of specimens played a role in the process of compiling a *pulu* but were later discarded in the publication process while the fourth section selects examples of painted *pulu* albums to reflect on the supposed advantage of these paintings over printed illustrations. The fifth and final section explores three examples of illustrated *pulu* depicting the multitude of varieties of a species and the challenges of tackling their similarities by means of illustrations alone.

1. Book Titles and Bibliographical Classification: Perspectives on *Pu* and *Tu*

There are several indicators for identifying a work as belonging to the *pulu* genre, namely, title, content, and structure, and, lastly, the authors' self-positioning within the tradition of *pulu* writing. The identification of the *pulu* genre, but also of the knowledge fields it constitutes, can further be guided by the classificatory assessment by bibliographers and library owners from the cultural sphere and the time of the respective works. This first section of the chapter investigates two of these aspects: firstly, how the choice of work titles reflects the presence or absence of illustrations; and secondly, how early bibliographical classification schemes sketched the relation of *tu* and *pu* titles to each other and to their specific knowledge area.

The title of a *pulu* in most cases simply combines the subject of the work (i.e., the specific species or artefact) with either the term *pu* or *lu*.[3] Whereas in cases such as 'chrysanthemum treatise' (*Ju pu*) the title reliably identifies the work to be of the *pulu* type, the appearance of *tu* (diagram/illustration) in a title does not guarantee that there are illustrations in the work, which also applies, of course, to works other than *pulu*. Two telling examples are the titles *Illustrated Eulogies of Remarkable Fish* by Yang Shen and *Detailed Register for a Book on Bamboo* (Zhupu xianglu; preface 1301) by Li Kan. Yang Shen's work, despite having 'illustrated' in its title, is plain text; Li Kan, however, heavily relies on a combination of text and image to describe his bamboo varieties, without the book title making any mention

3 *Pu* and *lu* are used as signifiers in about half of all *pulu* titles but appear together as *pulu* only as label of the bibliographical classification. For a more in-depth analysis of *pulu* titles, see Siebert, *Pulu – 'Abhandlungen und Auflistungen'*, pp. 135–146.

of it. Both titles and what their authors had to say about this discrepancy will be discussed in some detail below.

Two modern historians have used bibliographies of historical book titles to estimate the number of illustrated works. Dong Censhi counted 170 *pulu* titles dating to the Song dynasty (960–1279) of which forty-two contain illustrations – or at least say so in their title. Their larger share, i.e., twenty-eight titles, are works on antiques, coins, and seals; fourteen titles feature plants, tea, incense, and the scholar's studio as their topic.[4] This proportion of supposedly illustrated *pulu* declines significantly when taking those from pre-Song times until the end of the Qing dynasty (1644–1911) into account. Of the 1,189 *pulu* references I collected over the years, only sixty-three, i.e., five per cent, contain the term *tu* in their title. An even lower percentage of illustrated titles are found in Du Xinfu's *Bibliography of Ming Editions* (Mingdai banke zonglu). Craig Clunas counted only two hundred titles labelled as 'illustrated,' which amounts to two per cent of all titles listed in this bibliography. Clunas suggested two possible reasons for this, which seem to contrast with the popularity of illustrations during the Ming. One possible reason for this phenomenon, according to Clunas, is that elite scholars viewed books advertising their illustrations as objects of vulgarity and luxury; another one is that illustrations were possibly too common at the time to warrant special mention in the title.[5] Definitive statistics on illustrated publications of the Song, Ming, and Qing dynasties (including information on size and position of the illustrations within the book and on the page, etc.) will require a digital analysis of a large corpus of scanned images. However, the contributors to *Graphics and Text in the Production of Technical Knowledge in China* provided a thorough investigation of the functions of illustrations and diagrams in various textual traditions and fields of knowledge in imperial China.[6] Especially Bray's definition of *tu* as a functional rather than stylistic category, which can be applied to both monochromatic line drawings in printed *pulu* texts and to richly coloured images in painted albums, provides an analytical framework for this chapter.[7] Bray differentiates between two main types of illustrations: "transformative *tu*" aiming to lead the reader to spiritual advancement, and "representational, illustrative *tu*" conveying explanatory and instructional information. *Pulu* illustrations appear to mainly belong to the latter

4 Dong Censhi, "Min hui shi, xiu cheng pu, xie cheng tu – Songdai pulu zhong de tupu," *Wenjin xuezhi* (2018): 142.

5 Craig Clunas, *Pictures and Visuality in Early Modern China* (London: Reaktion Books, 1997), pp. 34–35. Clunas' statistic is based on Du Xinfu's catalogue, which lists nearly ten thousand titles; see Du Xinfu, *Mingdai banke zonglu* (Jiangsu: Jiangsu guangling guji keyin she, 1983).

6 See Francesca Bray, Vera Dorofeeva-Lichtmann, and George Métailié, eds., *Graphics and Text in the Production of Technical Knowledge in China: The Warp and the Weft* (Leiden: Brill, 2007).

7 Bray et al., *Graphics and Text*, p. xiii.

type, although they may also sometimes have been aimed to trigger a reader's inner sensation.[8]

Since the Song dynasty and the *Library Catalogue of the Suichu Hall* (Suichutang shumu, compiled in the twelfth century), the catalogue of the private library of You Mao (1127–1194), the stage was set for *pulu* writings to be assigned their own bibliographical classification.[9] The following short excursion widens the perspective slightly and follows some early bibliographical traces that suggest a special classificatory assessment of illustrated works (*tu*) and works of the treatise/register (*pu*) type. In these assessments, *pu* and *tu* were considered supplementary to each other and to their specific knowledge field. When considering the relation of *tu* and *pu*, the natural starting point is the Song dynasty philosopher Zheng Qiao, mentioned at the beginning of this chapter. Zheng stipulated the complementary roles of illustrations and texts to safeguard and transfer knowledge. His statement in the chapter "On illustrations and treatises" (Tupu lüe) of his encyclopaedic *Comprehensive Records* (Tongzhi, published 1161) counts as the *locus classicus* of this notion:[10]

> Illustrations are the warp and texts the weft; one warp, one weft, when woven together they become 'writing/culture' (*wen*).
>
> ...
>
> Reading the text but not looking at the illustration is like hearing the sound without seeing the shape; seeing the illustration without reading the text is like seeing the person but not hearing what he says.[11]

In his chapter "On bibliography" (Yiwen lüe) of the same work, Zheng Qiao exemplifies how he assessed the role of *tu* and *pu* type works of the past. Zheng subdivided several of his bibliographical classes into special sub-classes or sections for *tu*

8 Bray et al., *Graphics and Text*, pp. 33–34. A case of the 'transformative' type is surely the *Treatise on Mei-Plum Blossom Portraiture* (*Meihua xishen pu*) by Song Boren of the thirteenth century. Song depicts the various stages of the blossoming of Prunus *mume* (translated as either plum or apricot) and the positions of the flower on the twig as moral-ethic metaphor for the scholar. Maggie Bickford, *Bones of Jade, Soul of Ice: The Flowering Plum in Chinese Art* (New Haven, CT: Yale University Art Gallery, 1985). For a translation and reproduction of all illustrations, see Peter Wiedehage, *Das Meihua xishen pu des Song Boren aus dem 13. Jahrhundert: Ein Handbuch zur Aprikosenblüte in Bildern und Gedichten* (Nettetal: Steyler Verlag, 1995). However, Song's work is almost never identified as a *pulu* by Chinese bibliographers. In their view, it belonged to the bibliographical classification of 'art' (*yishu*).

9 For an account of how *pulu* have moved through the classification schemes see Siebert, *Pulu – 'Abhandlungen und Auflistungen'*.

10 Unless stated otherwise, all translations are mine.

11 Zheng Qiao, *Tongzhi*, in *Siku quanshu*, edited by Ji Yun et al. (Beijing, 18th century, reproduced in *Wenyuange Siku quanshu*, Shanghai: Shanghai guji chubanshe, 2003), vol. 372–381, *juan* 72, p. 1b.

and *pu* types of titles. A closer consideration of the hierarchical level of these *pu* and *tu* sub-classes reveals that their placement was determined by the field of knowledge they belonged to. In some fields, Zheng saw them as integral and he mixed *tu*/*pu* and non-*tu*/*pu* titles within the same class; in others, he counted them as supplementary but separate classes to the respective knowledge field. Zheng structured his bibliographical classification scheme into twelve main classes (*lei*) with two levels below, i.e., sub-classes (*xiao lei*) and sections (*mu*). Within the first two main classes, i.e., "Canonical writings" (*jing lei*) and "Rites" (*li lei*), several sub-classes have slots for *pu* and *tu* type works, but they appear only at the lowest level of the sections (*mu*).[12] In the main classes of "Medicine and therapy" (*yifang lei*) and "Five agents" (*wuxing lei*), however, Zheng installed them directly under the respective main classes as the sub-classes "Illustrations of *materia medica*" (*bencao tu*) and "Yijing diagrams" (*yi tu*). An example for sub-classes in which works with *tu* in their titles are not separated from the non-*tu* works is "Astronomy/Astrology" (*tianwen lei*). This suggests that Zheng Qiao considered illustrated works in this knowledge field as a standard and more integral part of writing, while in others, such as "Medicine" and "Rites," they constitute supplements and are – at different levels – peripheral to the more text-oriented canon of knowledge.

When we search for the already flourishing genre of *pulu* writings in Zheng Qiao's bibliography we find them already acknowledged as a separate classification. Zheng placed titles which would later fill the classification labelled *pulu* under the heading "economics" (*shihuo lei*, literally "Food and goods"), a sub-class under the main class "History" (*shi lei*).[13] This model was rarely followed by later bibliographers.

When looking further back into the history of bibliographical classification and a time when *pulu* were not yet established as a genre, we find two authors who like Zheng Qiao assessed *tu* and *pu* as supplemental categories to knowledge fields. Wang Jian (452–489) even introduced a separate "Tract of illustrations and treatises" (*tupu zhi*) as the seventh and last main category in his bibliography *Seven Tracts* (Qizhi). This work is no longer extant, but Ruan Xiaoxu (479–536), a bibliographer who saw himself as Wang's successor, reported about this special feature of Wang's classification scheme in his own bibliography, the *Seven Records*

12 Five of the sub-classes of the "Canonical writings" and four of the class "Rites" have sections plainly named *tu* or *pu*. Other sections add a specifier to their designation as for example the section "Illustrations of clothing" (*fu tu*) in the sub-class "Mourning garments" (*sang fu*) or the section "Genealogical treatises" (*shi pu*) in the sub-class "Annals of the Spring and Autumn Era" (*Chunqiu*). A diagram of all three levels of Zheng Qiao's bibliography is presented for example in Yao Mingda, *Zhongguo mulu xueshi* (Beijing: Shangwu yinshuguan, 1998, reprint of 1938 edition), pp. 108–110.

13 The sub-class "Food and goods" is further subdivided into the bibliographical sections (*mu*) of "Goods and valuables" (*huobao*), "Objects of use" (*qiyong*), "Animal care" (*huanyang*), "Plant cultivation" (*zhongyi*), "Tea" (*cha*), and "Wine" (*jiu*).

(Qilu). Nevertheless, Ruan dissolved this class again, and – as he explains in his preface – moved Wang's "illustrations and treatises" back to the end of the class or section they belong to according to their topic.[14] Like Zheng Qiao, Ruan treated *tu* and *pu* as appendices or supplements to classification groups defined by topic. Wang Jian, however, had considered *pu* and *tu* type works as supplementary to the whole set of the other six classifications in his bibliographical scheme.

Wang and Ruan addressed notions of *pu* and *tu* at a time when *pulu* was still developing as a genre. These traces of early bibliographical handling of *pu* and *tu* titles may point to one reason why *pulu* evolved as a separate scholarly genre, namely, the necessity to store and retrieve knowledge considered supplemental to the core of orthodox knowledge fields, such as the works Zheng Qiao put into the section "Illustrations of clothing" within the "Rites" classification (see footnote 12). The topics of *pulu*, however, quickly developed beyond those relevant for understanding canonical writings. In addition, *pulu* provided systematic information and detail not essential for contextualising a species or an object mentioned in the canon. Eventually, *pulu* established itself as a special form of writing at the periphery of acknowledged knowledge fields but by no means outside established forms of scholarly expertise. The supplementary position of *tu* and *pu* to the 'core' of scholarly production allowed *pu*-text and *tu*-image to function as equal and sometimes even interchangeable means of producing specialised knowledge. Therefore, whether *pulu* contained illustrations or not became a secondary issue. As will be discussed below, although they were interchangeable, scholars seem to have favoured text over image. The following sections of this chapter point out some of the issues that may have led to this preference.

2. *Pulu* without Illustrations

As already noted at the beginning of this chapter, most *pulu* rely on text alone to convey the 'idea' of the appearance and the character of a plant, animal, or object to their reader.[15] *Pulu* includes a broad range of available knowledge ranging from

14 Ruan Xiaoxu's bibliography is also lost today. Only his extensive preface is handed down by its inclusion into a collection compiled in the late seventh century. A copy of Ruan Xiaoxu's preface to his *Seven Records* can be found, for example, in Yuan Yuanqiu and Zeng Jiguang, eds., *Zhongguo lidai tushu zhulu wenxuan* (Beijing: Beijing daxue chubanshe, 1997), pp. 175–186. For Ruan's description of Wang Qian's bibliography and his rearrangement of the "Tract of illustrations and treatises," see the pages 177–178.

15 I use the word 'idea' here deliberately. Many descriptions use, to put it positively, a spacious verbal framework to outline a species, variety, or object. This is sometimes because the author himself has never really seen the 'thing' in question. However, what also may have played a role is the notion that the closer one looks, the more differences and varieties one detects. Interconnected communities, such

things seen or experienced by the author to textual sources. Authors of *pulu* often freely combined empirical knowledge and cited sources. However, the use of these different sources did not determine the absence or presence of illustrations in *pulu*. A good example for this is the illustrated *Treatise on Stones from Suyuan Garden* (Suyuan shipu; preface 1613) on decorative garden stones. The author, Lin Youlin (fl. early seventeenth century), depicted stones he had encountered during his travels and those he had only read about. In his preface, Lin admits having seen only about one fifth of the stones described and depicted in his work. "But the specialist can surmise the whole even when he has learned only about a part of it" – an attitude for which the compilers of the *Collected Abstracts of the Imperial Manuscript Library of All Four Bibliographical Classes* (Siku quanshu zongmu tiyao) criticised his work as being unreliable.[16]

Chrysanthemums provide a good example of *pulu* with (see section five of this chapter) and without illustrations. The plant was widely cultivated in people's gardens since the tenth century and became one of the most prominent topics of *pulu* writing. With over sixty known *pulu* titles, it is only outnumbered by *pulu* on tea and ink slabs. Chrysanthemums were popular as flowers with numerous shapes and colours and as a symbol of the scholarly hermit flourishing in the autumn of his life. The three non-illustrated *pulu* on chrysanthemums discussed below highlight the various considerations behind the absence of illustrations.

The first example is Liu Meng's (n.d.) *Treatise on Chrysanthemums* (Ju pu; preface 1104), one of the first *pulu* on chrysanthemums. Liu's work addressed several recurring topics in *pulu* on nature studies. He describes thirty-five chrysanthemum varieties, reflects on changes in the appearance of a flower generated by professional gardening, the regionality of their names and finally discusses the difficulty of correctly identifying varieties. His work is presented as the result of a visit to a scholar friend living in the north, far from Liu's home in Fujian. The friends had spent their time chatting about chrysanthemums and arranging the chrysanthemum collection of Liu's friends into a hierarchy based on the aesthetic appeal of their colour, fragrance, and "grace" (*tai*).[17] Liu Meng did not touch upon the subject of

as those established by Western scientists from the seventeenth century on, developed and agreed on basic anchors for identifying differences and identities between species; aficionado groups of orchid and chrysanthemum lovers – in China and everywhere else – tend to develop their own sets of defining criteria that are individual for the respective species and reflect the group's own focus of interests and enthusiasms.

16 Lin Youlin, *Suyuan Shi pu, fanli* ('Reading Guidelines'), p. 1a (reprint *Xuxiu Siku quanshu*, Shanghai: Shanghai guiji chubanshe, 1995–2002, vol. 1112, p. 605); Yong Rong, *Siku quanshu zongmu tiyao* (Beijing, eighteenth century, reprint, Beijing: Zhonghua shuju, 1987, 4th edition), vol. 116, p. 999.

17 Liu Meng, *Liu shi ju pu, pu xu* (Preface to the Treatise), in Ji Yun et al., eds., *Siku quanshu* (Beijing: eighteenth century, reproduced in *Wenyuange Siku quanshu*, vol. 845. Shanghai: Shanghai guji chubanshe, 2003), pp. 2a–b.

illustrations. And, indeed, the features he focused on were difficult to capture in an illustration. Moreover, Liu seems to have wanted the whole setting to be remembered: two learned gentlemen exploring various facets of chrysanthemums while interacting with the actual plants at his friend's home on that specific occasion.[18]

Most *pulu* authors pursued a different approach to Liu's event-driven account. They aimed at covering their chosen topic comprehensively, spending years to assemble their material. In addition to observing actual plants, they networked with likeminded friends and knowledgeable merchants, and located books on the topic by other authors. As a result, they faced a plethora of names and regional variations that were often impossible to match with varieties of their time and region. Adding to this the mutability of the individual appearance of a flower over time, some scholars decided to use only textual sources because they provided stable reference points and addressable units of knowledge.

An example of a seemingly purely philological project is Yatao's (active mid- to late-nineteenth-century) compilation of chrysanthemum names, *Chrysanthemum Treatise of the Huayun Balcony Parlor* (Huayunxuan ju pu, 1867). Probably overwhelmed by the 520 names he had assembled, Yatao chose an arrangement according to their rhyme category.[19] A more standard perspective of *pulu* writing on chrysanthemums is presented in the *Collected Treatises on Hundreds of Chrysanthemums* (Bai ju jipu) by Shi Zhu, compiled between 1242 and 1250.[20] In this work, Shi Zhu described thirty-nine chrysanthemum varieties he grew in his garden and referenced the content of five *pulu* on chrysanthemums from his library. He described the varieties from his garden when they were in a state of full bloom, because, as he stressed, it was necessary to know which stage of the flowering was described to identify a variety. Shi Zhu did not include any illustrations and seems

18 Liu does add some references to chrysanthemums he had not seen at his friend's home and thus does not include them in the hierarchy they had established together. One example is the "wild chrysanthemum" (*ye ju*), since Tao Hongjing (456–536) an icon for the life of a hermit, which does "preserve its 'as-is' [naturalness]" (*bao qi ziran*) and thus is an ideal a scholar should emulate. See Liu Meng, *Liu shi ju pu*, p. 19a.

19 Yatao's work, the *Chrysanthemum Treatise of the Huayun Balcony Parlour* (*Huayunxuan ju pu*), is included with an abstract in the *Xuxiu Siku quanshu zongmu tiyao* (*gaoben*) (Draft for a Sequence to the '*Siku quanshu*') (Beijing: Zhongguo kexueyuan tushuguan, 1996, vol. 5, pp. 307f); it is not included in the collection of the same name, the *Xuxiu Siku quanshu*. The work might not be extant anymore. Yatao was one of the sobriquets of Xu Baoqian (1817–1897) who seems to be the Yatao responsible for this chrysanthemum work. Xu is listed with this sobriquet as author of medical writings in Xue Qinglu, *Quanguo zhongyi tushu lianhe mulu* (Beijing: Zhongyi guji chubanshe, 1992), pp. 147, 273.

20 The two dates derive from the prefaces to the *Collected Treatises on Hundreds of Chrysanthemums* and its appendix, the *Supplement to the History of Chrysanthemum* (*Ju shi buyi*). The earliest extant printing of the work is of much later date and part of the collection *Miscellaneous Works from the Mountain Residence* (*Shanju zazhi*) of the Wanli era (1573–1620). This version of Shi Zhu's work is also the one used when compiling the late eighteenth century *Imperial Manuscript Library of All Four Bibliographical Classes* (*Siku quanshu*). Shi Zhu, *Bai ju jipu*, in *Siku quanshu*, vol. 845.

to have transferred what he saw directly into words. Moreover, it appears that he even excised the images that probably once belonged to the *Treatise Illustrating the Shapes of Chrysanthemums* (Tuxing ju pu), i.e., one of the five *pulu* he had added to his work.[21] Shi Zhu stated that he had "extracted the important parts" (*zhe qi yao*) of this work, which – as we know from Shi Zhu's compilation today – meant the preface and the names of forty-one chrysanthemum varieties but not the illustrations.[22]

3. Illustrations with Mediating Roles in the Process of Compiling *Pulu*

The examples above prioritised texts over illustrations for various reasons. In addition, in these examples, illustrations appear not to have played any role in the production process of the work. In the examples of *pulu* writings discussed in the following section, illustrations did play a role, either as incentives to produce a *pulu* or as tools to assemble a collection of items as the basis of a *pulu*. In both types, the illustrations were nevertheless not included in the published works. A telling example of the first type is the already mentioned *Illustrated Eulogies of Exceptional Fish*, which – despite its title – contained no illustrations. According to Yang Shen's 1544 preface, the inspiration to compile a work on this topic was indeed triggered by an illustrated work titled *Illustrations of Exceptional Fish* (Yiyu tu) compiled by an unknown author of the Southern Dynasties (420–589). Yang's original plan was to expand the *Illustrations of Exceptional Fish*. However, realising that it contained numerous errors, he turned to early "medieval treatises on local products (*wuzhi*) and other works" to compile his own version of a work on fish and other marine animals.[23] In terms of style, he decided to follow the model set by

21 The author of the *Treatise Illustrating the Shapes of Chrysanthemums* is given as Hu Rong, a native of Chicheng (Hebei). According to Shi Zhu, this work was finished in 1191 in originally two *juan* (roughly "volume") (Shi Zhu, *Bai ju jipu, juan* 5, p. 1a).

22 The only existing prints of Shi Zhu's compilation are those included into the book collection *Miscellaneous Works from the Mountain Residence* (footnote 20). In case he had not excised the images himself, the inclusion of Shi Zhu's work into this collection could have been the typical turning point in the history of the *Treatise Illustrating the Shapes of Chrysanthemums* in which illustrations were omitted due to cost constraints. Wang Yuhu's bibliography of Chinese agricultural writings lists Shi Zhu's work in a way that suggests an individual printing but does not give any specifics – and also does not mention any illustrations; Wang Yuhu, *Zhongguo nongxue shulu* (Shanghai: Nongye chubanshe, 1964), p. 99. However, none of the contemporary rare book catalogues supports the existence of such a separate printing. In all cases I have checked, so-called "Ming prints" turned out to be copies from the *Miscellaneous Works from the Mountain Residence* collection.

23 Yang Shen, *Yiyu tuzan*, in *Siku quanshu*, vol. 847, *yin* ('Lead-in'), p. 1a (reprint *Siku quanshu*, vol. 847). In the early seventeenth century, Yang, in turn, inspired another author. Hu Shi'an (1593–1663) continued Yang's work with a 'Supplement' (*Yiyu tuzan bu*) and 'Surplus collection' (~ *runji*) in 1618 and 'Commentaries'

Guo Pu's *Illustrated Eulogies on the 'Mountain and Sea Canon'* (Shanhai jing tuzan). Whereas Guo Pu's eulogies were composed to relate to illustrations, Yang skipped the illustrations but kept the name "illustrated eulogy" (*tuzan*). In Yang's view, the reader could envision the fish based on words alone, so there was "no need to use colourful images to show off their splendour".[24] The Qing dynasty bibliographer Zhou Zhongfu (1768–1831) interpreted this as the replacement of illustrations with eulogies.[25]

The term "illustrated eulogy" (*tuzan*) was a genre marker for works in the tradition of Guo Pu's style, in which a eulogy (*zan*) highlights one's appreciation of the illustration or the illustrated (*tu*).[26] The term *tuzan* was later often used for portraits of the author of a book or the key characters in a novel which together with a eulogy of the depicted were added at the beginning of a work. Yang Shen's decision to call his work an "illustrated eulogy" despite omitting illustrations was an exceptional move, one that was continued by Hu Shi'an in the first half of the seventeenth century (see footnote 23). We can assume that Yang chose the style of a eulogy to provide information in a condensed and scholarly acceptable format. For his own work, he considered his words sufficient and illustrations unnecessary for passing on relevant information. Two centuries later, the compilers of the *Imperial Manuscript Library of All Four Bibliographical Classes* took issue with this assertion and criticised Yang for omitting illustrations.[27]

Another type of involvement of illustrations in the production of a non-illustrated *pulu* is exemplified by a *pulu* on lychee (*Litchi chinensis*) by the scholar-official

(~*jian*) in 1630. Hu kept the term "illustrated eulogies" in the title but also did not add any illustrations (reprint *Siku quanshu*, vol. 847, pp. 751–848). The painted album Yang refers to in his preface is lost and no information about the author or illustrator is known. Moreover, the title is quite generic. At the end of the nineteenth century, Zhao Zhiqian (1829–1884) painted a two-metre-long scroll of the same title with depictions of various fish and seafood arranged together with their names and descriptions (see contribution by Wang Ching-ling, fig. 6.7). Fish books closer to the style of *pulu* are the illustrated album *Illustrated Seafood* [*Album*] (Haicuo tu, published as *Qinggong Haicuo tu* by the Gugong chubanshe in 2014), and the fish album discussed by Wang Ching-ling in this volume.

24 Yang Shen, *Yiyu tuzan, yin*, p. 1a.

25 Zhou Zhongfu, *Zhengtang dushu ji* (reproduced in *Xuxiu Siku quanshu*, Shanghai: Shanghai guji chubanshe, 1995-2003), vol. 924–925, *juan* 51, p. 16b. Zhou's use of the word *dai* ("replace/stand in") confirms my reading of Yang's phrase as discarding illustrations. Others have interpreted Yang's phase as an opposition to a "flowery and incomprehensible" language; Yang Zhao, "Yang Shen 'Yiyu tuzan' de wentixue yiyi," *Jianghan luntan* 2 (2012): 101–104.

26 Dong Censhi strictly differentiates between *tupu*, "illustrated treatises," and *tuzan*, "illustrated eulogies." While in his interpretation the first mostly follow the path of an objective display supplementing the verbal description, the latter are a means to express a personal, subjective evaluation in rhymed verse of the object or person depicted. See Dong, "Min hui shi, xiu cheng pu, xie cheng tu," 146. Roel Sterckx also briefly addresses *tuzan*; Sterckx, "The Limits of Illustration: Animalia and Pharmacopeia from Guo Pu to *Bencao Gangmu*," *Asian Medicine* 4 (2008): 357–394, especially 365f.

27 Yong Rong ed., *Siku quanshu zongmu tiyao*, vol. 115, p. 996.

Cai Xiang (1012–1067). Cai had served as prefect in several locations in Fujian and Zhejiang, where lychee fruits were abundant. He thus decided to dedicate the first *pulu* to this topic. Cai reports that each time he encountered a lychee of superior quality, he evaluated its features and asked a professional painter to draw it "from life" (*ming gong xiesheng*).[28] Having assembled a sufficient number of images and notes, he finally sat down to compile his *pulu*, inaugurating lychee into the body of *pulu*. However, the images of the fruits had only served as visual aids in his study of the topic. He discarded them in the final publication.[29]

It is unclear how often images were used as aids in the process of producing *pulu*. The previous chapter by Ching-Ling Wang implies that this might have happened more frequently than the imageless printed versions we see today suggest. At least one more example of a printed *pulu* with a production workflow like Cai's can be named. Zhang Qiande (1577–1643), or Zhang Chou as he later called himself, entertained a lifelong passion for goldfish. He had encountered many different specimens and had the finest ones painted (*tuxie*) by professional painters. In contrast to Cai, Zhang did not produce individual entries on the varieties based on the paintings but presented more abstract types of goldfish deduced from recurring colour patterns and shapes in his work.[30]

As will be discussed in the last section, adding illustrations to a *pulu* could involve intellectual and technical challenges. And there was always, of course, the fundamental challenge of increased printing costs. Authors also considered the

28 Cai Xiang, "Yuan benshi" (Exploring [the Topic's] Beginning and History), in *Lizhi pu*, in *Siku quanshu*, vol. 845, p. 154. "*Xiesheng*" later became one of the early Chinese terms for "photography."

29 From a letter to Cai Xiang by another famous Song-dynasty author, Ouyang Xiu (1007–1072), we learn that Ouyang had made copies of "lychee illustrations" (*lizhi tu*) owned by Cai. Ouyang's letter dates to 1065, that is six years after Cai Xiang had finished his *pulu* on lychee and two years after Ouyang had written a preface to the work. Dong takes this time discrepancy to imply that the illustrations are not the ones that served as the basis for Cai Xiang's book. Dong, "Min hui shi, xiu cheng pu, xie cheng tu," p. 150.

30 Zhang Qiande, *Zhushayu pu* (*Treatise on Vermillion Fish*), in *Meishu congshu* (Taibei: Guangwen shuju, reprint 1963), pp. 125–134; the above quote can be found on page 126. A.C. Moule published an extensive article including a translation of this work under the title "A version of the Book of Vermillion Fish," *T'oung Pao* 39 (1950): 1–82. The same year – and in correspondence with Moule – George Hervey had published *The Goldfish of China in the XVIII Century* (London: The China Society, 1950). His volume contains two translations of materials relevant to the history of knowledge about Chinese goldfish in Europe. Firstly, the first European book on goldfish, the illustrated *Histoire naturelle des dorades de la Chine* by Edme-Louis Billardon de Sauvigny published in Paris 1780, and secondly, a *Mémoire* sent to Henri-Leonard Bertin (1720–1792) a few years before Sauvigny's book by Jesuit missionaries in China together with a painted scroll, six metres in length, depicting ninety-two goldfish. The history and depictions of Chinese goldfish in the West is another topic. The fish dramatically changed their appearance when copied from the Chinese scroll into Sauvigny's book. They had lost their bulging eyes and veil-like tails and appear just like interestingly coloured carps. The original goldfish scroll is kept at the library of the Muséum National d'Histoire Naturelle in Paris. Sauvigny, *Histoire naturelle des dorades de la Chine* (Paris, 1780, digital copy, Harvard University, Museum of Comparative Zoology, Ernst Mayr Library, https://doi.org/10.5962/bhl.title.6902).

accessibility of their work, which might suffer when illustrated or even painted. This argument was put forth by Kong Wuzhong (1042–1097) in a critique of his predecessor Liu Ban's (1022–1088) writing on herbaceous peonies (*shaoyao*, i.e., *Paeonia lactiflora*). Liu Ban favoured painted albums to distribute his knowledge about the plants. He claimed to have commissioned illustrations for his work to "more fully inform those unfamiliar with the plant, and to prove the reliability of his account to those who already knew it."[31] A few years later, Kong criticised this elitist attitude in his own *Treatise on Herbaceous Peonies* (Shaoyao pu), stating that "only those of our generation with sufficient resources can fetch a skilled painter to provide illustrations [of their varieties of herbaceous peonies], flaunting them everywhere. Nothing surpasses a description in writing to make [knowledge] universally accessible to all."[32] Paradoxically, both works did not survive as individual works. Liu's manuscript encountered the fate of numerous illustrated works as only its preface and the names of the herbaceous peonies have survived in an encyclopaedic collection of the thirteenth century. Despite Kong's argument on the accessibility of his *pulu*, his work only survived as a section in a twelfth-century collection of brush notes.

Yang Shen and Kong Wuzhong considered textual descriptions as substitutes for or equivalents to illustrations of objects. In exceptional instances, *pulu* authors thought the text of a *pulu* could also function as an 'illustration' in its own right. The *Classic of Tea* (Cha jing) by Lu Yu (733–804), an early model text for *pulu* writing, covers standard *pulu* subjects such as the history of tea, its manufacture, and tea-producing areas. The tenth and last chapter of the *Classic of Tea* is labelled "Illustration of tea" (*Cha zhi tu*), although it does not contain any visual representations. Instead, Lu advised his readers to transcribe the preceding nine chapters onto four or six pieces of silk and to display them in the tearoom. This would allow the text "to catch the eye and be instantly memorable" (*ji mu er cun*).[33] Cai Xiang's *Record of Tea* (Cha lu, epilogue dated 1064) probably envisioned a similar use of text as image. Cai had this text carved in stone to preserve its legacy in more permanent form, allowing later generations to take rubbings from the stone which could similarly be turned into an item for display.[34]

31 Liu Ban, *Shaoyao hua pu*, reproduced in Zhu Mu (Thirteenth century), *Gujin shiwen leiju houji* (Nanjing, 1604/1763), *juan* 30, pp. 16a–18a. The quote translated above is found on page 17a (see digitised copy provided by Bavarian State Library, https://www.digitale-sammlungen.de/view/bsb00059658?page=229).

32 Kong Wuzhong, *Shaoyao pu*, reproduced in Wu Ceng (twelfth century), *Nenggaizhai manlu*, in *Siku quanshu*, vol. 850, *juan* 15, p. 21b.

33 Lu Yu, *Cha jing*, in *Siku quanshu*, vol. 844, *juan xia*, p. 14b.

34 Cai Xiang, *Cha lu*, in *Siku quanshu*, vol. 844, p. 5b. The fate of this stone is reported in a Ming dynasty epilogue to the *Cha lu* by Xu Bo (1570-1642). Xu Bo, *Chong bian Hongyulou tiba* (seventeenth century, reproduced in *Shumu congbian*, 3rd collection, Taibei: Guangwen shuju, 1969), vol. 50, *juan* 1, 25b–26a.

4. *Pulu* and painting

Inventories of the paintings stored at the Qing palaces prove that painted albums of flowers and animals were abundant. Some of them might have shown an affinity to *pulu* writing, with texts attached that would still make a complete work even when printed without the illustrations. The following is not meant as a comprehensive reflection on the genre of painted albums of nature studies topics; instead, it draws attention to some of the features of these painted albums and their relation to *pulu* writing.

Painted albums like that of Liu Ban on herbaceous peonies were exclusive objects for presentation and personal use. Producing and reproducing them required an expert's hand. They thus rarely circulated widely or left a lasting legacy among the scholarly community, at least when compared to works like Lu Yu's *Classic of Tea*, which influenced centuries of *pulu* scholars' writing on tea and other *pulu* topics.[35] Of Liu's knowledge of peonies, only a list of names and a preface remain. The set of illustrated albums commissioned by the Qianlong Emperor – which has recently gained attention among historians – was even more exclusive.[36] Whereas the colourful illustrations paired with descriptive texts in the albums on birds and beasts would have met *pulu* readers' expectations, the *Book of Doves* (Boge pu) presents a paradigm like that of Liu Ban. This album only gives the names of the variety of dove depicted. Moreover, the names are separated from the painting and seem replaceable as they are written on small yellow paper slips, glued onto the frame of the painting. The focus of the *Book of Doves* seems twofold. On the one hand, it registers the morphological features of the dove while on the other hand animating it with lively postures showing its character.[37] According to Lai

35 P. F. Kornicki tells a different story of manuscripts being copied and distributed in Japan in the seventeenth century, with some of them also including illustrations. Kornicki, "Manuscript, Not Print: Scribal Culture in the Edo Period," *Journal of Japanese Studies* 32, no. 1 (2006): 23–52. In a similar way the manual copying of *pulu* texts for personal use – from printed but also from manuscript exemplars – surely was a central driving force for the distribution of the specialised knowledge they contain. But when it comes to copying painted albums, I figure, other more elitist goals might have played a role. I have not investigated how Ouyang Xiu made or wanted to make use of his copies of Cai Xiang's lychee paintings (see footnote 29). However, the visual content of painted albums, as discussed above, did not distribute that easy or widely as those consisting of printed text only.

36 I here refer to the *Book of Birds* (Niao pu), *Book of Beasts* (Shou pu) and *Book of Doves* (Boge pu) published in colour in 2014 by the Beijing Palace Museum as *Qinggong Niao pu* etc., and the online exhibition of several painted albums at the Taiwan Palace Museum (https://theme.npm.edu.tw/exh106/manualofbird/ch/index.html). The most in-depth analysis on the *Book of Birds* and *Book of Beasts* has been published by the Taiwanese scholar Lai Yu-chih. See for example her "Images, Knowledge and Empire: Depicting Cassowaries in the Qing Court," *Transcultural Studies* 1 (2013): 7–100.

37 See for example the image of courting doves of the "Chrysanthemum phoenix" (*Juhua feng*) variety, in *Book of Doves* (Boge pu), ascribed to Jiang Tingxi (1669–1732) (rep. Beijing: Beijing Palace Museum Publishing House, 2014), no. 10 (see also https://photo.sina.cn/album_26_17348_27830.htm).

Yu-chih, this style became somewhat outdated in the high Qing and was replaced by a new style, represented by albums such as the *Book of Birds* (Niao pu, 1761) which depicts birds in stiff poses in the extreme foreground, suggesting a more object-focused approach.[38] In the *Book of Birds*, the colourful illustrations and the accompanying descriptive texts received equal scholarly input by the authors. Even without the images the texts would still have provided sufficient knowledge of the species to satisfy the needs of contemporary scholars, and they would have made a complete *pulu*.

Like the *Book of Birds*, Zou Yigui's (1668–1772) album of twenty-four varieties of peonies was commissioned by Emperor Qianlong. However, his peonies are depicted as plucked flowers against an empty background, detached from the plant and its environment. Each flower is accompanied by a short description. With the image of a flower on the front and its description on the back of the folded album leaf, both types of information occupy the same amount of space in the album, putting equal emphasis on image and description.[39] Again, even when losing the images, the descriptions alone would still have constituted a complete *pulu*.

These painted albums – like the preparatory work by Cai Xiang for his *pulu* on lychee – document that the relation between visual and textual discourse had been close in some cases. The *Detailed Register for a Book on Bamboo* by Li Kan, a painter famous for paintings of bamboo, is another interesting case in this respect. The first part of his work is in the style of a painter's manual and explains how to draw the various parts, settings, and life stages of bamboo. The second part, however, is in the style of a *pulu* and assembles more than three hundred bamboo varieties including plants that "resemble bamboo but are not" (*si shi er fei zhu pin*), each with a short description and a simple drawing of the plant.[40] In the first part

38 Lai, "Images, Knowledge and Empire," pp. 50–51. Lai takes the *Book of Birds* as an example of images that reference a "text-external real world, not the established textual tradition … [and thus] convey much information that is not, or could not be, encoded in text" (Lai, "Images, Knowledge and Empire," p. 56). For my interpretation of the *Book of Birds* and the multifaceted relationship of image, description, naming, and poetry to produce 'consumable' nature, see Siebert, "Consuming and Possessing Things on Paper". For an online presentation of the image of the common crane (*Xiao he xian*) discussed there see *Collection of National Palace Museum, Taibei* by the Taiwan Palace Museum (http://theme.npm.edu.tw/exh106/manualofbird/common/images/selection/img1_1_7.jpg). The *Book of Birds* was conjointly authored by Yu Sheng (1692–1767) and Zhang Weibang (eighteenth century) on the basis of an album by Zhang Tingxi (1667–1732).

39 This album is part of a collection of painted albums at the National Palace Museum, Taiwan. Digital images of the album leaves can be found by following this search link: https://digitalarchive.npm.gov.tw/List/Index. In 1756, ten years after the peony album, Zou Yigui's produced for the Qianlong emperor an album on "foreign [i.e. Japanese] chrysanthemums" (*yang ju*). The album is lost but may have been similar to the above album. For more details on Zou's work on chrysanthemums see Siebert, "Consuming and Possessing Things on Paper," pp. 392–396.

40 In the seven-*juan* (roughly "volume") version of the *Detailed Register for a Book on Bamboo* used here, this second part takes up the last five *juan*.

on drawing bamboo, Li Kan states that one precondition to painting bamboo is to know its names and understand its various shapes and forms.[41] In his view, painting bamboo requires specialist knowledge derived from *pulu*. The second part of the book, on the other hand, rarely addresses matters of painting. In a few places the author notes that some bamboo varieties were popular painting subjects but does not give specific examples. Possibly due to Li Kan being a well-known painter, Chinese historical bibliographers classified the book in the 'art' category based on its first part on drawing bamboo.[42]

Painted albums present image and text as separate but at the same time mutually supportive entities. In painting scrolls the boundary between painting and *pulu* text sometime becomes blurred. Examples such as the *Scroll On the Local Products from Ouzhong* (Ouzhong wuchan juan) and *Images of Exceptional Fish* (Qiyu tu) combined description and image into one visual experience.[43]

5. Depicting a multitude of varieties

The albums with colour illustrations of plants and animals discussed above also added pleasure to reading. Their illustrations convey the idea or the ideal of the plant or animal while the use of colour and the three-dimensional appearance elicit – as Georges Métailié puts it – "a deeper feeling of a living plant [or animal]." At first glance, they appear to be best at depicting minute differences between varieties of a species. However, as Métailié puts it, albums are not necessarily "more precise than pictures in technical books" such as those in *materia medica* (*bencao*), which served as mnemonic devices for those already familiar with the item in question.[44] While *materia medica* captured mainly the iconic appearance of a species and only rarely included regional or other varieties, *pulu* aimed at including a multitude of varieties of a species; they were geared towards a different mode of close-up scholarly engagement with their subjects. For the expert, they might have similarly functioned as mnemonic devices, although the illustrations of the varieties of a species were unavoidably repetitive.

41 Li Kan, *Zhu pu xianglu*, in *Zhibuzu zhai congshu*, edited by Bao Tingbo, in *Baibu congshu jicheng*, no. 29 (Taipei: Yiwen yinshuguan, 1965–1970), *juan* 2, p. 1a.

42 One rare example of classifying the work as *pulu* is the *Book Catalogue of the Jingu Hall* (Jingutang shumu) dated to either late Ming or early Qing. Anonymous compiled, *Jingutang shumu*, in *Yujian zhai congshu*, edited by Luo Zhenyu (tome 14. S.l., 1910), *shang*, p. 36b.

43 On these two scrolls, see Anran Tu, *Pictures for Action: Painting and Collecting Nature in Modern China, from Zhao Zhiqian (1829–1884) to Jin Cheng (1878–1926)* (MA thesis, University of Alberta, 2018).

44 George Métailié, "The Representation of Plants: Engravings and Paintings," in *Graphics and Text*, pp. 493, 498.

The following section will discuss three distinguished examples of *pulu* depicting large sets of varieties of a species in print, with each author approaching the challenge of repetitiveness and unavoidable similarity of the images differently.

The first example is the *Chrysanthemum Book of Deshan's Hermitage* (Deshanzhai ju pu) by Zhu Youkuang (d. 1471), published in 1458. The work presents one hundred chrysanthemum varieties, each including a simple illustration of the flower, stem, and leaves, accompanied by a poem with the variety's name as title and a short inline comment describing its main morphological features.[45] Text and illustration are printed on one piece of paper, but after being folded and bound into book form, they cannot be viewed side by side. Therefore, the reader must turn over the text page to see the illustration. This was necessary for keeping the two types of information securely together in the process of printing and binding, but also significantly shaped the reading experience. Figures 7.1a-d reproduce text and illustration of the first four chrysanthemum varieties in Zhu Youkuang's book. The depiction of the varieties appears repetitive, seemingly showing little concern for guiding readers to help them differentiate between the varieties. However, readers would encounter the illustrations only after first reading name, inline comment and poem and were therefore prepared for 'reading' the illustrations. The first variety in Zhu's book is the "Golden Peacock" (see figure 7.1a). Inline comment and poem roughly translate as:

> *Golden Peacock*
> (thousand-petal variety of deep-yellow colour with vermillion centre)
> As if golden fur and jade-green plumage come flying,
> forming blossoming flowers of autumn splendour.
> A Southern demeanour is still present,
> and stirred by the wind they appear like flying birds.

Names and short inline comments taught readers about key features and helped them differentiate between varieties. These two pieces of information also appear at the beginning of Zhu Youkuang's book in the table of contents. The illustrations and poem added liveliness and imaginative power to Zhu Youkuang's packaging of knowledge about chrysanthemum flowers.[46]

45 Zhu Youkuang, *Deshanzhai ju pu*, reproduced in *Meiguo Hafo Yanjing tushuguang cang zhongwen shanben huikan*, series 1, vol. 27 (Beijing: Shangwu yinshuguan, 2003). A digital copy is available at Harvard Digital Library: https://iiif.lib.harvard.edu/manifests/view/drs:54074139.

46 I have not done a systematic comparison between the printed chrysanthemums of *Deshan's Hermitage* and those painted in Zou Yigui's album mentioned above (footnote 39). What appears striking when comparing the variant "Ten thousand volume book" (*Wanjuanshu*) that appears in both works, is, that first the images differ significantly and second, while the painted album gives a more neutral description

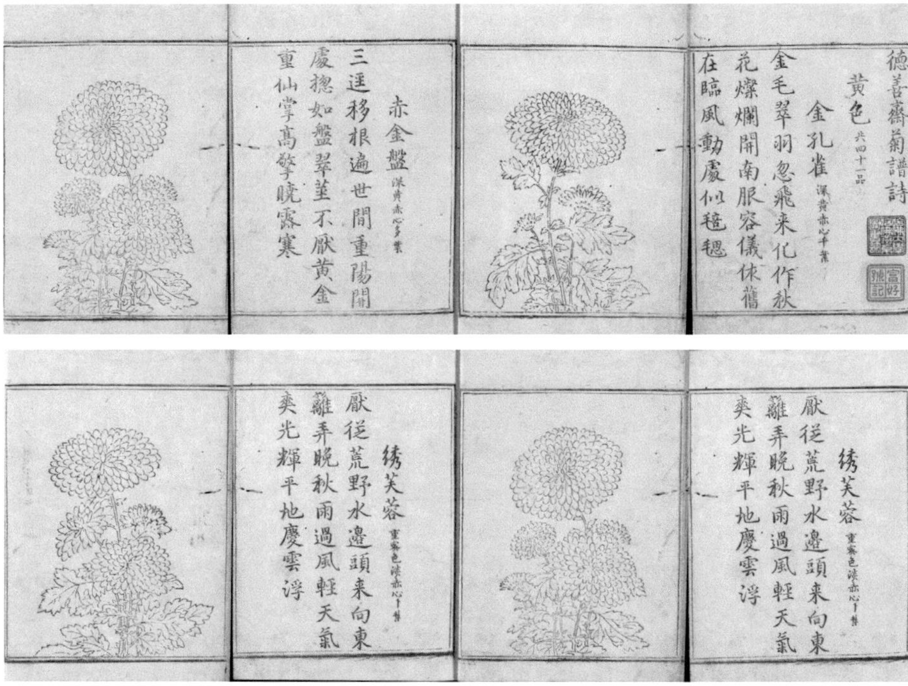

Figure 7.1 (from right to left): the first four chrysanthemum varieties of the *Deshanzhai Jupu* (1458) by Zhu Youkuang (d. 1471). Harvard College Library Harvard-Yenching Library (https://iiif.lib.harvard.edu/manifests/view/drs:54074139).

The second example, *Illustrated Treatise on Goldfish* (Jinyu tupu, preface 1848), depicts fifty-four varieties of goldfish in three-colour printing. According to the author who called himself "The hermit farmer from Mount Juqu" (Juqu shannong), the illustrations were based on a painted album of goldfish by a certain Mr. Xue, the only dedicated book on the subject known to him. Aware of the lack of information in Xue's book, the author decided to compile a more instructive *pulu* including practical information on goldfish breeding. To popularise Xue's album, he reproduced a printed version as an appendix to his work, but the illustrations have no added text and even the names of the varieties need to be deduced from the table of contents preceding the appendix. Different from the *Chrysanthemum Book of Deshan's Hermitage*, the *Illustrated Treatise on Goldfish* puts the illustration on the front side of the page, leaving the back side empty, as if waiting to be inscribed with a description or a poem by a future owner of the copy (Figures 7.2a–d). Only two illustrations come with published text, namely, the two variants at the end

of shape and naming conventions, the print adds a poem instead. See https://digitalarchive.npm.gov.tw/Painting/Content?pid=13027&Dept=P# and https://nrs.lib.harvard.edu/urn-3:fhcl:27733728?n=50.

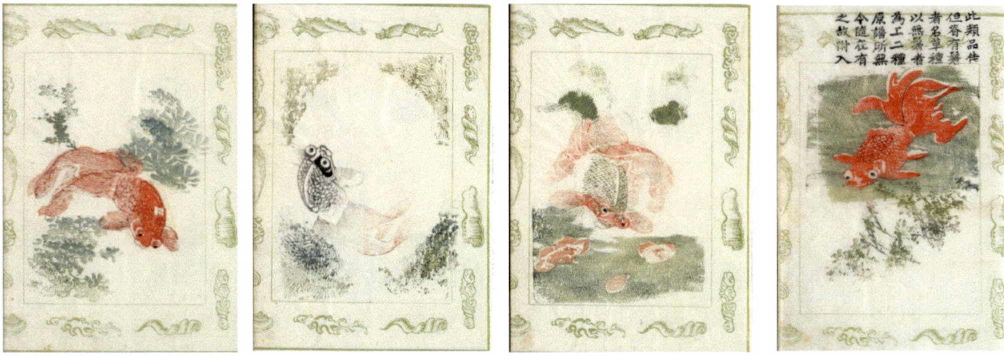

Figure 7.2 (from left to right): a-c) first three goldfish varieties depicted in the appendix to the *Jinyu tupu* ("Illustrated Treatise on Goldfish," preface 1848); d) the second "newly added ordinary variety". *Jinyu tupu, bu fen juan*, copy from National Library of China, World Digital Library. (https://www.loc.gov/resource/gdcwdl.wdl_17217/?sp=14 to https://www.loc.gov/resource/gdcwdl.wdl_17217/?sp=16, and https://www.loc.gov/resource/gdcwdl.wdl_17217/?sp=69).

of the appendix labelled newly added "ordinary types" (*fan pin*) in the table of contents. However, these texts do not occupy the empty verso of the page but are printed as marginalia onto the decorative frame and above the image of the fish itself, leaving the back of the page empty again (see Figure 7.2d).[47]

The author of the *Illustrated Treatise on Goldfish* chose the more versatile technique of colour printing for the illustrations. Yet, even with the addition of colour it would have been difficult to identify, for instance, a real variety on sale at the market using one of these illustrations. Nevertheless, the images surely added attractiveness to the small booklet for buyers, showing them the possible shapes and patterns of a multitude of goldfish varieties. The separation of Xue's illustrations from the "new" *pulu* may have been caused by the intention to transmit the "old" *pulu* in an uncorrupted way. The cumbersome aligning of the names in the table of contents with their respective images inadvertently conveyed to readers how easily the non-professional could drown in the swirl of goldfish varieties.

The third example of depicting varieties turns to the orchid, one of the most popular *pulu* topics of the late Qing dynasty; orchids had long competed with chrysanthemums for scholarly attention. The *Record of Being One Heart with Orchids*

47 That the book is not some sort of unfinished project we can conjecture from the existence of at least two copies still extant today at the National Library of China in Beijing. The *World Digital Library* reproduces one of these copies, see https://hdl.loc.gov/loc.wdl/wdl.17217. The *Illustrated Treatise on Goldfish* is not the first *pulu* on goldfish. However, both preceding ones only circulated in manuscript form. They are first the *Treatise on Vermillion Fish* written 1596 (see footnote 30) and the *Treatise on Cinnabar [Coloured] Fish* (*Zhuyu pu*) written by Jiang Zaiyong in 1699 (reprint *Xuxiu Siku quanshu congshu*, vol. 1120). Some of the depictions of the *Illustrated Treatise on Goldfish* can be matched with the descriptions in Jiang's work. See Siebert, "Animals as Text," pp. 155–156.

(Lanhui tongxin lu, published in 1891) by Xu Naihe (1834–1911) is one of the few *pulu* on orchids containing illustrations. Xu's work presents the plant on a par with calligraphy and as a topic equally worthy for scholars. The work is divided into three relatively distinct parts of roughly the same size. It starts with prefaces and dedications written in various calligraphic styles, followed by descriptions of morphology and pedigree, poems, and illustrations of fifty-eight varieties of orchids,[48] to finally elaborate on the criteria for evaluating these varieties and giving advice on the care of the plant. Xu explains that he produced the illustrations by "tracing the shadow [of flower and stem] cast by a lamp" (*bi ying goule*) to capture the individual shape of each orchid variety.[49] Figure 7.3b gives examples of how he assembled text and illustration into one visual arrangement on the page. Matching the technique of tracing the shadow of the plant, Xu also copied the calligraphic styles in the first part of his book by "tracing the originals" (*shuanggou*). In so doing, he treated the shapes of the orchid flowers and the shapes of the calligraphic styles in similar fashion, emphasising the idea of orchids as a scholarly plant.

In the book's "Guide to the reader" (*fanli*), Xu foregrounded his identity as a connoisseur and as a new type of orchid lover who prioritised the quality of the flower over the number of varieties. He thereby declares his work to be different from *pulu* written during the early phase of the genre, i.e. the Song dynasty.[50] The emphasis on authorship and the mode by which he approached the topic are also evident in the two portraits of himself and his father Xu Yunqin (n.d.) at the beginning of the *Record of Being One Heart with Orchids*. While his father is holding an orchid flower as a sign of the practical engagement of the Xu family with the topic, Xu Naihe is holding a book to demonstrate the scholarly foundations of the task (Figure 7.4). The images underscore the author's conviction that a perfect *pulu* needed to combine both aspects.

48 For two examples of this, see Figure 7.3a–b. The descriptions in the *Lanhui tongxin lu* put a focus on tracing the history or pedigree of a variety, namely, who brought it when to what city first. Image of Figure 7.3b is an especially elaborate example of this.

49 Ji Nan (1760–1834), used a similar technique of light and shadow, to emphasise the shape of his flowers on display during on the evenings of chrysanthemum spectacles for his guests. Ji Nan, *Ju shuo*, in *Zhaodai congshu*, compiled by Zhang Chao (nineteenth century, reproduced, Shanghai: Shanghai guji chubanshe, 1990), p. 2020.3.

50 Xu Naihe, *fanli* ('Guide to the reader'), in *Lanhui tongxin lu* (Jingfang xianguan edition, 1891), n.p. A digital copy is available at Harvard Yenching Digital Library via Hathi Trust: https://catalog.hathitrust.org/Record/100369850. The societal aspect of publishing one's own work played its part also for *pulu* authors. Giving one's name and engagement as a scholar to a book on a topic that is neither canonical, nor moral, nor of relevance to the state's well-being, has always been something a *pulu* author might feel the need to defend himself. See Siebert *Pulu – 'Abhandlungen und Auflistungen'*, chapter 3.2.

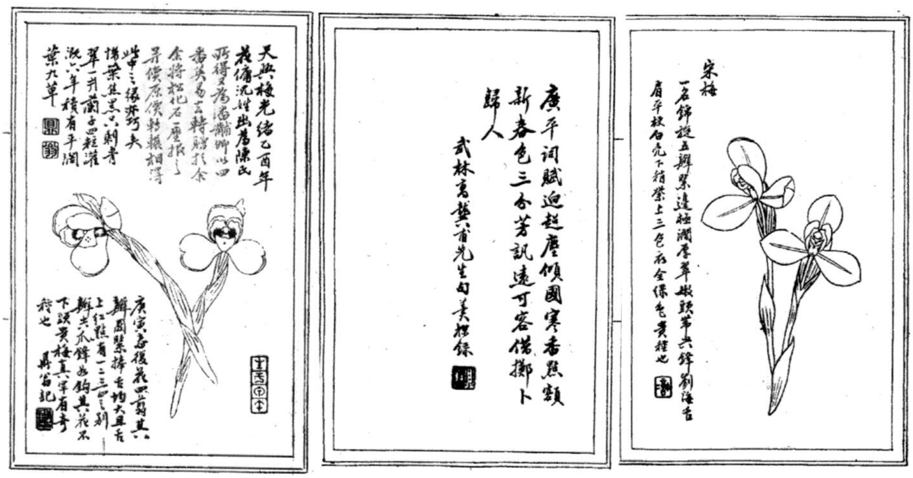

Figure 7.3 (from right to left): a) description and appreciative poem to the *Song mei* orchid variety; b) genealogy of the orchid variety *Tianxing mei*. Harvard College Library Harvard-Yenching Library, Public Domain (https://babel.hathitrust.org/cgi/pt?id=hvd.32044070481189&view=1up&seq=115 to seq=116 and https://babel.hathitrust.org/cgi/pt?id=hvd.32044070481189&view=1up&seq=147).

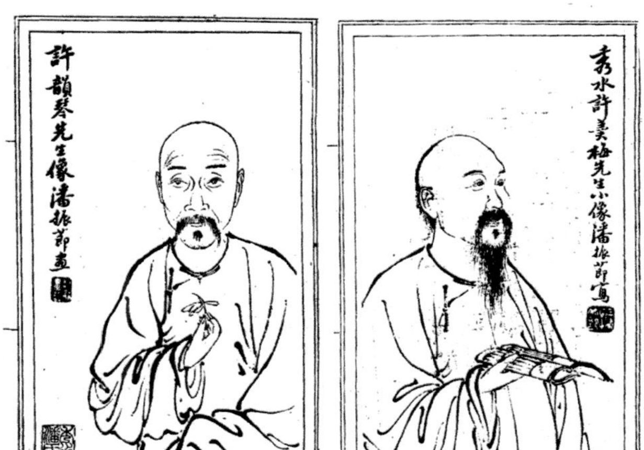

Figure 7.4. To the left: Xu Naihe, here given with his courtesy name Meimei; to the right: his father Xu Yunqin. Harvard-Yenching Library, Public Domain (https://babel.hathitrust.org/cgi/pt?id=hvd.32044070481189&view=1up&seq=29 and seq=31).

Conclusion

This chapter has explored the presence and absence of illustrations in *pulu* writing. In sum, non-illustrated *pulu* were dominant as many authors seemed not to consider illustrations – or illustrations alone – as sufficient to fully describe a species or variant. Text in the form of names, descriptions, and poetry was needed to supplement illustrations and bring them to life. In the examples above, text could function as a stand-in or at least an equivalent to illustrations in *pulu* writing. In other cases, authors used paintings of specimens in the process of *pulu* writing to research the topic and to draft the text before discarding them in the final print version. At the same time, there were also very practical reasons for the absence of illustrations, such as increased publication costs or considerations regarding the difficulty of reprinting illustrated texts. Some *pulu* originally had illustrations but later lost them in the process of reprinting. Since one of the defining purposes of *pulu* was to serve as comprehensive collections of varieties of a species, illustrations were especially challenging; they needed to capture the intricacies of the differences between varieties of a certain animal or plant. Equally challenging was the task of elucidating through painted or printed illustrations some of the features considered relevant in the scholarly community of flower or goldfish lovers, such as fragrance or agility, their pedigree or the author's emotional connection to the species. Most *pulu* known today have no illustrations, mainly because authors considered text as the central and most reliable mode for packaging and conveying knowledge and the most stable form for its transmission. When *pulu* were quoted in other contexts – in regional descriptions, *materia medica*, or in works on 'superfluous things'[51] in scholarly lives – illustrations were not missed. The *Collected Abstracts of the Imperial Manuscript Library of All Four Bibliographical Classes* contains several exceptions to this. The work criticises Yang Shen for overestimating the descriptive quality of his words on fish and for discarding the images in his *Illustrated Eulogies of Remarkable Fish* (footnote 27). It also blames the compiler of the thirteenth-century collection *The Sea of Learning Fed by Hundreds of Rivers* (Baichuan xuehai) for leaving out the illustrations when including Tang Ji's work on inkstones.[52] Other abstracts rarely push for including images. Although we can only conjecture, readers of *pulu* seem to have agreed with this viewpoint.

The short excursion in the first section of this chapter highlighted traces of a special treatment of illustrated (*tu*) and register-like (*pu*) works in Chinese bibliographical classification schemes of the fifth and sixth centuries, that is, before *pulu* writings appeared as a genre. These classification schemes considered *tu* und *pu* type works as supplementary to the core of library holdings. The notion of *tu* and *pu* acting on the same level and as supplementary to the core of traditional scholarship was developed further by Zheng Qiao in the twelfth century. The connection between *tu* and *pu* – I conjecture – has been one of the major driving forces behind the development of *pulu* writing from the tenth century onwards. With

51 I borrow this term from Clunas' book on a genre of guide books to elite living that also drew material from *pulu* writing. See Clunas, *Superfluous Things: Material Culture and Social Status in Early Modern China* (Honolulu: University of Hawai'i press, 2004).

52 See abstract on the *Illustrated Treatise on the Inkstones from She County* (She yan tupu, colophon 1066) by Tang Ji (eleventh century) in *Siku quanshu zongmu tiyao*, vol. 115, p. 984.

the specialisation of *pulu* contents in the multitude and the variability of species on the one hand and in the care, history, and poetry related to it on the other hand, *pu* became the favoured mode over *tu* for transferring and packaging knowledge.

I have excluded from this discussion questions about the images' 'scientificity' or the adaptation of any form of Western style 'accuracy' in the illustrations.[53] Even works like the *Small History of the Orchid* (Lanhui xiaoshi) by Wu Enyuan (1872–ca. 1932), published in 1923 and beyond the scope of this volume, does not venture too far outside the frame set by earlier *pulu* writings despite using photography to depict orchid varieties. Wu wrote the variety's name and the description directly onto the photograph and around the flower itself in a visual arrangement strikingly like the one used by Xu Naihe.[54]

My focus on the examples of illustrated *pulu* introduced above was not so much their success in tackling the challenge of depicting variety, but to understand what the authors might have wanted to achieve with these images or what seems to have been important to them. We must assume that Zhu Youkuang's drawings of chrysanthemums accompanied by short inline comments and poems adequately expressed the 'idea' of chrysanthemum variants in his view and most probably that of his readers as well. Xu Naihe produced imitations of calligraphy styles and shapes of orchid variants using a similar tracing technique for both, giving us an idea about how he might have perceived both. Like calligraphy styles based on examples by famous predecessors, Xu's orchid variants arose from legacy. And this was probably what both text and illustrations were intended to achieve: to give an idea of the species or variety in question and, at the same time, of the author, his engagement with the topic and his belonging to a tradition of like-minded scholars. The author-portraits added to Xu's work give additional weight to this idea.

Glossary

Bai ju jipu	百菊集譜
Baichuan xuehai	百川學海
bao qi ziran	保其自然
bencao tu	本草圖

53 Problems arising from this type of questions have been investigated by several scholars. Frederico Marcon's *The Knowledge of Nature* (Chicago, IL: University of Chicago Press, 2015) and Maki Fukuoka's *The Premise of Fidelity* (Stanford: Stanford University Press, 2012), both relate to Japan but are most instructive on the *problematique* of authenticity and scientificity of images and knowledge representation in general.

54 These *Lanhui xiaoshi* illustrations are examples of visual objects characterised by Yi Gu as "[blurring] the boundary between photography and other pictorial media." See Gu Yi, "What's in a Name? Photography and the Reinvention of Visual Truth in China, 1840–1911," *The Art Bulletin* 95.1 (2013): 123.

bi ying goule	逼影鉤勒
Boge pu	鵓鴿譜
Cai Xiang	蔡襄
cha	茶
Cha jing	茶經
Cha lu	茶錄
Cha zhi tu	茶之圖
Chicheng	赤城
Chunqiu	春秋
dai	代
Deshanzhai ju pu	德善齋菊譜
Du Xinfu	杜信孚
fan pin	凡品
fanli	凡例
fu tu	服圖
Guo Pu	郭璞
Hu Rong	胡融
huanyang	豢養
Huayunxuan ju pu	華韻軒菊譜
huobao	貨寶
ji mu er cun	擊目而存
Ji Nan	計楠
Jiang Tingxi	蔣廷錫
Jiang Zaiyong	蔣在雝
Jin kongque	金孔雀
Jingutang shumu	近古堂書目
jing lei	經類
Jinyu tupu	金魚圖譜
jiu	酒
Juhua feng	菊花鳳
Ju pu	菊譜
Ju shi buyi	菊史補遺
Juqu shannong	句曲山農
Kong Wuzhong	孔武仲
Lanhui tongxin lu	蘭蕙同心錄
Lanhui xiaoshi	蘭蕙小史
lei	類
Li Kan	李衎
li lei	禮類
Lin Youlin	林有麟

Liu Ban	劉攽
Liu Meng	劉蒙
lizhi tu	荔枝圖
Lu Yu	陸羽
Meihua xishen pu	梅花喜神譜
Meimei	美梅
ming gong xie sheng	命工寫生
Mingdai banke zonglu	明代版刻綜錄
mu	目
mudan	牡丹
Niao pu	鳥譜
Ouyang Xiu	歐陽修
Ouzhong wuchan juan	甌中物產卷
pu xu	譜敘
pulu	譜錄
Qilu	七錄
Qiyu tu	奇魚圖
qiyong	器用
Qizhi	七志
Ruan Xiaoxu	阮孝緒
sang fu	喪服
Shanhai jing tuzan	山海經圖讚
Shanju zazhi	山居雜誌
shaoyao	芍藥
Shaoyao hua pu	芍藥花譜
Shaoyao pu	芍藥譜
She yan tupu	歙硯圖譜
shi lei	史類
shi pu	世譜
Shi Zhu	史鑄
shihuo lei	食貨類
Shou pu	獸譜
shuanggou	雙鉤
Siku quanshu	四庫全書
si shi er fei zhu pin	似是而非竹品
Song mei	宋梅
Song Boren	宋伯仁
Suichutang shumu	遂初堂書目
Suyuan shipu	素園石譜
tai	態

Tang Ji	唐積
Tao Hongjing	陶宏景
tianwen lei	天文類
Tiangxing mei	天興梅
Tongzhi	通志
tupu	圖譜
Tupu lüe	圖譜略
tupu zhi	圖譜志
tuxie	圖寫
Tuxing ju pu	圖形菊譜
Wang Jian	王儉
Wu Enyuan	吳恩元
wuxing lei	五行類
wuzhi	物志
Xiao hui he	小灰鶴
xiao lei	小類
Xu Baoqian	徐寶謙
Xu Naihe	許鼐龢
Xu Yunqin	許韻琴
Xue	薛
Xue shi jiu pu	薛氏舊譜
Yatao	亞陶
yang ju	洋菊
Yang Shen	楊慎
ye ju	野菊
Yi tu	易圖
yifang lei	醫方類
yin	引
yishu	藝術
Yiwen lüe	藝文略
Yiyu tu	異魚圖
Yiyu tuzan	異魚圖贊
yuan	源
Yuan benshi	原本始
zan	贊
Zhang Chou	張丑
Zhang Qiande	張謙德
Zhang Tingxi	張廷錫
Zhao Zhiqian	趙之謙

zhe qi yao	摭其要
Zheng Qiao	鄭樵
zhongyi	種藝
Zhu pu	竹譜
Zhu Youkuang	朱有爌
Zhupu xianglu	竹譜詳錄
Zhushayu pu	朱砂魚譜
Zhuyu pu	朱魚譜
ziyu zhi guo	自詡之過
Zou Yigui	鄒一桂

Bibliography

Primary Sources

Cai Xiang 蔡襄. *Lizhi pu* 荔枝譜, in Ji Yun 紀昀 et al. eds. *Siku quanshu* 四庫全書. Beijing: 18th century. Reproduced in *Wenyuange Siku quanshu* 文淵閣四庫全書 845. Shanghai: Shanghai guji chubanshe, 2003.

Cai Xiang 蔡襄. *Cha lu* 茶錄, in Ji Yun 紀昀 et al. eds. *Siku quanshu* 四庫全書, Beijing: 18th century. Reproduced in *Wenyuange Siku quanshu* 文淵閣四庫全書 844. Shanghai: Shanghai guji chubanshe, 2003.

Hu Shi'an 胡世安. *Yiyu tuzan bu, ~ runji, ~ jian* 異魚圖贊補, ~閏集, ~箋, in Ji Yun 紀昀 et al. eds. *Siku quanshu* 四庫全書, Beijing: 18th century. Reproduced in *Wenyuange Siku quanshu* 文淵閣四庫全書 847. Shanghai: Shanghai guji chubanshe, 2003.

Ji Nan 計楠. *Ju shuo* 菊說. In *Zhaodai congshu* 昭代叢書, compiled by Zhang Chao 張潮. S.l., 19th century. Reproduced, Shanghai: Shanghai guji chubanshe, 1990.

Jiang Tingxi 蔣廷錫 (attributed). *Boge pu* 鵓鴿譜. Beijing: 18th century. Reproduced, Beijing: Beijing Palace Museum Publishing House, 2014.

Jiang Zaiyong 蔣在雍. *Zhuyu pu* 朱魚譜. S.l., 1699. Reproduced in *Xuxiu Siku quanshu congshu*, vol. 1120. Shanghai: Shanghai guji chubanshe, 1995-2003.

Jingutang shumu 近古堂書目, compiled by anon., in *Yujian zhai congshu* 玉簡齋叢書, Luo Zhenyu ed. 羅振玉, tome 14. S.l., 1910.

Juqu shannong 句曲山農. *Jinyu tupu* 金魚圖譜. S.l., 1848. Digital copy, World Digital Library, https://hdl.loc.gov/loc.wdl/wdl.17217, visited July 2023.

Li Kan 李衎. *Zhu pu xianglu* 竹譜詳錄, in Bao Tingbo 鮑廷博 ed. *Zhibuzu zhai congshu* 知不足齋叢書 10. Reproduced in *Baibu congshu jicheng* 百部叢書集成 29. Taipei: Yiwen yinshuguan, 1965–1970.

Lin Youlin 林有麟. *Suyuan Shi pu* 素園石譜. S.l, 1613. Reproduced in *Xuxiu Siku quanshu* 續修四庫全書 1112. Shanghai: Shanghai guji chubanshe, 1995-2003.

Liu Meng 劉蒙. *Liu shi ju pu* 劉氏菊譜. S.l., 1104, in Ji Yun 紀昀 et al. eds. *Siku quanshu* 四庫全書, Beijing: 18th century. Reproduced in *Wenyuange Siku quanshu* 文淵閣四庫全書 845. Shanghai: Shanghai guji chubanshe, 2003.

Lu Yu 陸羽, *Cha jing* 茶經. S.l., 8th century, in Ji Yun 紀昀 et al. eds. *Siku quanshu* 四庫全書, Beijing: 18th century. Reproduced in *Wenyuange Siku quanshu* 文淵閣四庫全書 844. Shanghai: Shanghai guji chubanshe, 2003.

Qinding Niao pu 欽定鳥譜. Beijing, 18th century. Reproduced in *Xuxiu Siku quanshu congshu* 續修四庫全書 1119. Shanghai: Shanghai guji chubanshe, 1995–2003.

Qinggong Niao pu 清宮鳥譜 = Catalogue of Birds Collected in the Qing Palace, Yu Sheng and Zhang Weibang. Beijing, 1761. Li Shi 李湜 and Gugong bowuyuan 故宮博物院 eds. Beijing: Gugong chubanshe, 2014.

Qinggong Shou pu 清宫兽谱 = Catalogue of Animals Collected in the Qing palace, Yu Sheng and Zhang Weibang. Beijing, 1761. Li Shi 李湜 and Gugong bowuyuan 故宮博物院 eds. Beijing: Gugong chubanshe, 2014.

Qinggong Haicuo tu 清宮海錯圖, anon. Beijing, 18th century. Wen Jinxiang 文金祥 and Gugong bowuyuan 故宮博物院 eds. Beijing: Gugong chubanshe, 2014.

Sauvigny, Edme-Louis Billardon de. *Histoire naturelle des dorades de la Chine*. Paris, 1780. Digital copy, Harvard University, Museum of Comparative Zoology, Ernst Mayr Library, https://doi.org/10.5962/bhl.title.6902.

Shi Zhu 史鑄. "*Bai ju jipu* 百菊集譜," in Ji Yun 紀昀 et al. eds., *Siku quanshu* 四庫全書. Beijing, 18th century. Reproduced in *Wenyuange Siku quanshu* 文淵閣四庫全書 845. Shanghai: Shanghai guji chubanshe, 2003.

Wu Ceng 吳曾. "*Nenggaizhai manlu* 能改齋漫錄 (12th century)," in Ji Yun 紀昀 et al. eds., *Siku quanshu* 四庫全書. Beijing, 18th century. Reproduced in *Wenyuange Siku quanshu* 文淵閣四庫全書 850. Shanghai: Shanghai guji chubanshe, 2003.

Wu Enyuan 吳恩元. *Lanhui xiaoshi* 蘭蕙小史. Shanghai: Zhonghua shuju, 1923.

Xu Bo 徐𤊹. *Chong bian Hongyulou tiba* 重編紅雨樓題跋 (17th century). Reproduced in *Shumu congbian* 書目叢編, 3rd collection, 50. Taibei: Guangwen shuju, 1969.

Xu Naihe 許鼐穌. *Lanhui tongxin lu* 蘭蕙同心錄. S.l.: Jingfang xianguan 竟芳仙館, 1891. Harvard Yenching Digital Library.

Yang Shen 楊慎. "*Yiyu tuzan* 異魚圖贊," in Ji Yun 紀昀 et al. eds. *Siku quanshu* 四庫全書. Beijing, 18th century. Reproduced in *Wenyuange Siku quanshu* 文淵閣四庫全書 vol. 847. Shanghai: Shanghai guji chubanshe, 2003.

Yong Rong 永瑢 ed. *Siku quanshu zongmu tiyao* 四庫全書存目提要. Beijing, 18th century. Reprint, Beijing: Zhonghua shuju, 1987 (4th edition).

Zhang Qiande 張謙德. "*Zhushayu pu* 朱砂魚譜 (1596)," in Hong Binhong 黃濱虹 et al. eds., *Meishu congshu* 美術叢書, series 2, vol. 10. Reprint, Taibei: Guangwen shuju, 1963.

Zheng Qiao 鄭樵. "*Tongzhi* 通志 (1161)," in Ji Yun 紀昀 et al. eds., *Siku quanshu* 四庫全書. Beijing, 18th century. Reproduced in *Wenyuange Siku quanshu* 文淵閣四庫全書 372–381. Shanghai: Shanghai guji chubanshe, 2003.

Zhongguo kexueyuan tushuguan 中國科學院圖書館 comp. *Xuxiu Siku quanshu zongmu tiyao* (*gaoben*) 續修四庫全總目提要 (稿本). Jinan: Qilu shushe, 1996.

Zhou Zhongfu 周中孚. *Zhengtang dushu ji* 鄭堂讀書記 (18th century). Reproduced in *Xuxiu Siku quanshu* 924–925, Shanghai: Shanghai guji chubanshe, 1995–2003.

Zhu Mu 祝穆. *Gujin shiwen leiju houji* 古今事文類聚後集. Nanjing, 1604/1763. Bavarian State Library – Digitale Sammlungen.

Zhu Youkuang 朱有爌. *Deshanzhai ju pu* 德善齋菊譜. S.l., Ming (1465–1620). Reproduced in *Meiguo Hafo Yanjing tushuguang cang zhongwen shanben huikan* 美国哈佛燕京圖書館藏中文善本彙刊, series 1, vol. 27. Beijing: Shangwu yinshuguan, 2003. Harvard Digital Library.

Secondary Sources

Bickford, Maggie. *Bones of Jade, Soul of Ice: The Flowering Plum in Chinese Art*. New Haven, CT: Yale University Art Gallery, 1985.

Bray, Francesca, Vera Dorofeeva-Lichtmann, and George Métailié eds. *Graphics and Text in the Production of Technical Knowledge in China: The Warp and the Weft*. Leiden: Brill, 2007.

Clunas, Craig. *Pictures and Visuality in Early Modern China*. London: Reaktion Books, 1997.

Clunas, Craig. *Superfluous Things: Material Culture and Social Status in Early Modern China*. Honolulu: University of Hawai'i press, 2004.

Dong Censhi 董岑仕. "Min hui shi, xiu cheng pu, xie cheng tu 一 Songdai pulu zhong de tupu 繁華事，修成譜，寫成圖 —— 宋代譜録中的圖譜," *Wenjin xuezhi* 文津學志 (2018): 142–153.

Du Xinfu 杜信孚. *Mingdai banke zonglu* 明代版刻綜錄. Jiangsu: Jiangsu guangling guji keyin she, 1983.

Fukuoka, Maki. *The Premise of Fidelity: Science, Visuality, and Representing the Real in Nineteenth-Century Japan*. Stanford, CA: Stanford University Press, 2012.

Gu, Yi. "What's in a Name? Photography and the Reinvention of Visual Truth in China, 1840–1911," *The Art Bulletin* 95.1 (2013): 120–138.

Hervey, George. *The Goldfish of China in the XVIII Century*. London: The China Society, 1950.

Kornicki, P. F., "Manuscript, Not Print: Scribal Culture in the Edo Period," *Journal of Japanese Studies* 32.1 (2006): 23–52.

Lai Yu-chih. "Images, Knowledge and Empire: Depicting Cassowaries in the Qing Court," *Transcultural Studies* 1 (2013): 7–100.

Marcon, Frederico. *The Knowledge of Nature and the Nature of Knowledge in Early Modern Japan*. Chicago, IL: University of Chicago Press, 2015.

Métailié, Georges. "The Representation of Plants: Engravings and Paintings," in Francesca Bray et al., eds., *Graphics and Text in the Production of Technical Knowledge in China: The Warp and the Weft*. Leiden: Brill, 2007, pp. 485–520.

Moule, A. C. "A Version of the Book of Vermillion Fish," *T'oung Pao* 39 (1950): 1–82.

Siebert, Martina. "Animals as Text: Producing and Consuming 'Text-Animals'," in Martina Siebert, Roel Sterckx, D. Schäfer, eds., *Animals Through Chinese History*. Cambridge: Cambridge University Press, 2018, pp. 139-159.

Siebert, Martina. "Consuming and Possessing Things on Paper: Examples from Late Imperial China's Natural Studies," in Elif Akçetin and Suraiya Faroqhi, eds., *Living the Good Life: Consumption in the Qing and Ottoman Empires of the Eighteenth*. Leiden and Boston, MA, Brill, 2017, pp. 384–408.

Siebert, Martina. "Neue Formen für neue Themen – *pulu* 譜錄 als bibliographische Kategorie und als Schriften zu Sach- und Naturkunde," in Florian C. Reiter ed., *Das Reich der Mitte – in Mitte. Studien Berliner Sinologen*. Wiesbaden: Harrassowitz, 2006, pp. 59–70.

Siebert, Martina. *Pulu – 'Abhandlungen und Auflistungen' zu materieller Kultur und Naturkunde im traditionellen China*. Wiesbaden: Harrassowitz, 2006.

Sterckx, Roel. "The Limits of Illustration: Animalia and Pharmacopeia from Guo Pu to *Bencao Gangmu*," *Asian Medicine* 4 (2008): 357–394.

Tu, Anran. *Pictures for Action: Painting and Collecting Nature in Modern China, from Zhao Zhiqian (1829-1884) to Jin Cheng (1878-1926)*. Master's thesis, University of Alberta, 2018.

Wang Yuhu 王毓瑚. *Zhongguo nongxue shulu* 中國農學書錄. Shanghai: Nongye chubanshe 1964.

Wiedehage, Peter. *Das* Meihua xishen pu *des Song Boren aus dem 13. Jahrhundert: ein Handbuch zur Aprikosenblüte in Bildern und Gedichten*. Nettetal: Steyler Verlag, 1995.

Xue Qinglu 薛清錄. *Quanguo zhongyi tushu lianhe mulu* 全國中醫圖書聯合目錄. Beijing: Zhongyi guji chubanshe, 1991.

Yang Zhao 楊釗. "Yang Shen 'Yiyu tuzan' de wentixue yiyi 楊慎 '異魚圖贊' 的文体學意義," *Jianghan luntan* 江漢論壇 2 (2012): 101–104.

Yao Mingda 姚名達, *Zhongguo mulu xueshi* 中國目錄學史. Beijing: Shangwu yinshuguan, 1998 (reprint of 1938 edition).

Yuan Yuanqiu 袁詠秋 and Zeng Jiguang 曾季光 eds. *Zhongguo lidai tushu zhulu wenxuan* 中國歷代圖書著錄文選. Beijing: Beijing daxue chubanshe, 1997.

About the Author

Martina Siebert is an independent scholar who also works as digital curator and area specialist for China at the Staatsbibliothek zu Berlin. She researches and publishes on scholarly explorations into nature and technology in imperial China with a focus on the styles and agendas of presenting and organizing that knowledge in writing.

Index

For Product Safety Concerns and Information please contact our EU representative GPSR@taylorandfrancis.com Taylor & Francis Verlag GmbH, Kaufingerstraße 24, 80331 München, Germany

Batch number: 10753435

Printed by Printforce, the Netherlands